emergency SANDBAG SHELTER
AND ECO-VILLAGE

MANUAL - HOW TO BUILD YOUR OWN
WITH SUPERADOBE / EARTHBAG

NADER KHALILI

COMPILED AND EDITED
ILIONA OUTRAM

Cal-Earth Press
Hesperia, California, USA

"earth turns to gold
in the hands of the wise"

- Rumi

Cal-Earth Press is the publishing wing of Cal-Earth Institute.

Cal-Earth (California Institute of Earth Art and Architecture) is dedicated to research and education into the universal elements of Earth, Water, Air and Fire, their unity in architecture, philosophy, and poetry in the service of the environment, arts and humanity.

Books by the same author:

Racing Alone
Ceramic Houses and Earth Architecture
Sidewalks on the Moon
Rumi, Fountain of Fire
Rumi, Dancing the Flame

Apprenticeship Film Series:

Making of a Dream
Earth Turns to Gold
Designing with Nature
Natural Materials
Eco-Dome, Building a Small Home

Front cover photograph: Sustainable village of prototype emergency shelters, Cal-Earth Institute, California.
Back cover photographs: top, architect Nader Khalili: bottom, compiler and editor Iliona Outram.

Persian calligraphy by Ali Heidari.

Structural Principles by Phill Vittore of P.J. Vittore Ltd.

Building Standards article reprinted from the September - October 1998 issue, with permission of the International Conference of Building Officials (I.C.B.O.)

EMERGENCY SANDBAG SHELTER
AND ECO-VILLAGE
Copyright © 2008 by E. Nader Khalili
All Rights Reserved
Printed in the United States of America

No part of this book may be used or reproduced in any manner whatsoever without written permission except in the case of brief quotations embodied in critical articles or reviews.
Exception: "EMERGENCY SANDBAG SHELTER: Quick Training Guide" on pages 19-22 may be copied and distributed free of charge for emergency shelters among the homeless and refugees of natural and man-made disasters.

For information address: Cal-Earth Press
10177 Baldy Lane
Hesperia, CA 92345
U.S.A.

website: **www.calearth.org**

FIRST EDITION 2008

Library of Congress Catalog Card Number: 2007943917

ISBN: 978-1-889625-05-8

This book is dedicated to empowering people to use the materials of war, sandbags and barbed wire, to make safe shelters.

"Bypass the U.N., bypass government. People can start building their own."

CONTENTS

	Contents	6
	Introduction	9
	Acknowledgements	11
CHAPTER 1	**SANDBAG SHELTER**	13
	Quick Training Guide	19
	A SINGLE DOME	23
	Layout and Orientation	32
	Foundation	35
	Base Wall	46
	Doorway	50
	Dome	52
	Windows	56
	Buttress Wall	68
	The Upper Dome	72
	Entry Vault	80
	Waterproofing and Finishing	85
CHAPTER 2	**ECO-VILLAGE AND MASS CLUSTERING**	87
	Design and Technique Variations	
	MODELS 2 & 3: Pouches	95
	MODEL 4: Pottery Dome	107
	MODEL 5: Koala Pouch	115
	MODEL 6: Holey Dome	127
	MODEL 8: Sinapsoapsis - Caterpillar.	143
	MODEL 9: Homeless Deluxe	149
	MODEL 10: Seashell Dome	179
	MODEL 1: Roofless Dome	193
	THE COURTYARD: Landscaping	203

CHAPTER 3	**SHELTER PROJECT EXAMPLES** U.N., UNIDO, UNHCR, NGO and Private	219
CHAPTER 4	**CHILDREN AND THE FUTURE** MODEL 7: Eco-Dome (Moon Cocoon)	233 238
APPENDIX		
I	**STRUCTURAL PRINCIPLES**	241
II	**ARCHES**	255
III	**SMALL BAGS**	267
IV	**COMPASS**	279
	Glossary **Chronology** **Index**	291 295 297

*"first plant your seed
then rely on the almighty"*

- Rumi

INTRODUCTION

To begin building an emergency shelter you don't need to read this manual entirely, but go to the "Quick Training Guide" in the next several pages and get started.

This book is one of the volumes in a series we have been preparing on Earth Architecture since 1998. The accelerating rate of disasters in recent years around the globe, natural and man-made, and the overwhelming requests for information have created a sense of urgency not to wait for the final manual, but to offer this book independently, and immediately. We therefore first summarized this book into four pages and put it on our website for instant and free global access as a Quick Guide for Emergency Sandbag Shelter. Not only has this move been received with great enthusiasm but it also generated a great many questions, mainly about climatic conditions, and how communities or eco-villages could be created.

Although we have tried to answer some of these questions in this volume we have also reached the conclusion that this book will always need updating to include new innovative techniques of quicker and more efficient building methods. The ever expanding applications of the Superadobe/sandbag/earthbag technology, both in building structures and infrastructure, and its flexible design possibilities, lend itself to a universal application ranging from a small emergency shelter to planetary habitat based on the utilization of local resources, some of which will be covered in the other volumes.

If a lack of uniformity is seen in the text and photos, it is because the material of this book has been taken from many sources such as lectures, workshops, and interviews given by Nader Khalili, and alternate construction techniques tried at Cal-Earth or other sites. We are showing a large number of photographs, so that pictures alone can speak to people who cannot read. We also show photos of imperfect, unprofessional, primitive examples built by inexperienced students, because our work has never been about staging a perfect final product, but it is about empowering people to build without being intimidated by thinking they are not skilled. Everything in earth architecture is beautiful. Being imperfect but safe is also O.K., it is even more human.

One of our main focus areas since the establishment of the Cal-Earth prototype site in 1991, has been championing the process of working very closely with the Building and Safety code requirements, local and national, as well as informing the building officials before we do everything we do, to gain their approval and respect. This alone can be considered a great success in the world of alternative and sustainable architecture. Many years of construction, officially approved structural analysis, testing and inspection, workshops and lectures most of which were recorded, have been studied to extract these pages, which represent variations in shelter design individually or clustered to make a village.

It is planned that the next volumes in this series on Earth Architecture will cover:

1) "Eco-Dome, a small home": how-to build your own, including part-blueprints and structural calculations.
2) "Superadobe Vaulted Houses": how-to build your own, including part-blueprints and structural calculations for:
a) "3-Vault House": a minimum standard house for a family according to the United Nations and
b) "Earth One": a standard 3-4 bedroom, 2 bath, 2-car garage house, and 3-Vault design variations.
3) "The Philosophy of Earth Architecture": covering the Timeless Materials of the Universal Elements, and Timeless Principles of design, using arches, vaults, and domes.
4) Infrastructure, Landscaping and Cityscaping.
5) Construction Documents, Structural Engineering and Building Codes.
6) Examples of building around the globe.

This simple technology was introduced at the first NASA sponsored Symposium for "Lunar Bases and Space Activities of the 21st Century", and later presented to the Los Alamos National Laboratory, amongst others. It has been endorsed and used by the United Nations, UNDP (United Nations Development Program), UNHCR (United Nations High Commission for Refugees), and UNITAR (United Nations Institute for Training and Research).

Building permits in California from 1994 and updated to 2008, were all to establish the credibility of a universal system of Earth Architecture in general and Superadobe in particular.

Many thousands have come to Cal-Earth Institute (a nonprofit organization) to learn in short workshops or long-term apprenticeships. They have invested many days and often a small fortune to travel and learn, and then gone around the world building and teaching. Yet more people are facing the difficulty of the heavy expenses of travelling to the United States. They have urged us to put all the knowledge into books and films, and to use electronic media to make this available globally.

Therefore this book and a number of films and documents are now available to empower these builders, supported by distance learning from Cal-Earth and other sources.

Superadobe/sandbag technology has been patented in the United States and overseas, to protect the innovator's right to make it freely available to the needy of the world and the owner-builder, and to license it for commercial use.

> To honor all those who have supported this work, we must add the following note:
> Although we wish this simple technology to be spead and reach people around the globe, we must also caution the users to verify the credibility of their source of information and its updated safety advice.

one by one
our friends
filled with joy and quest
begin to arrive

- Rumi

ACKNOWLEDGEMENTS

Many hundreds have participated - professionals, apprentices, associates, volunteers and students - to bring the Superadobe technology to this stage, and we are ever more thankful for their genuine contributions.

Although Cal-Earth has always insisted on being a self-reliant nonprofit organization, it has also been endowed by generous donations and grants from like-minded humanitarian foundations and individuals, such as the National Endowment for the Arts, the Ted Turner Foundation, Laura Huxley's Our Ultimate Investment Foundation, the Rex Foundation (Grateful Dead), Kit Tremaine, the Leventis Foundation, the Flora Family Foundation, and the Ashianeh Foundation. Recently, thanks to Barbara Michael, the Joseph Plan Foundation's donation has supported the free distribution of many copies of this volume.

We would like to thank all those who have contributed their time, money, inspiration, and hard work to begin the first volume of this series on Earth Architecture.

Special acknowledgment for the progress of these works goes to Phill Vittore of P.J. Vittore Ltd. whose vast knowledge and years of experience building large scale shell structures of domes and vaults, and the time which he gave generously as he created structural calculations, helped the process of successful testing of the prototypes for the seismic standards of California's building codes, one of the strictest in the world. We also acknowledge Carl Howe who put his stamp of approval on the first structural engineering plans submission.

Of the many apprentices, volunteers, and students who spent many years, or months alongside us at Cal-Earth, special thanks go to dear James Manzi, Farzin Soheili, Alex Nikraftar, Scott Kuhlman, Jann Rucquois, Michael Huskey, Kelly and Sky Sisson, Mark Harmon, Hooman Fazly, Ian Lodge and Michael Wood. We would like to acknowledge Khurram Schroff and Shahid Malik Shahid of Cal-Earth Pakistan, Behyiar and Heidi Ghahremani, the late Ali Akbar Khorramshahi, Dastan Khalili for spending hundreds of hours trekking from Los Angeles to Cal-Earth and creating films on this work assisted by Sheefteh Khalili, Ray Kappe the founder of SCI-Arc (Southern California Institute of Architecture) and the many architecture students since 1991, Madhu Thangavelu for planetary activities, Manijeh Mirdamad, Amir Farman-Farma, Afsaneh Bassirpour, Nassrine Alikhani and Theodore Petroulas, Nassrine Azimi, Anton Ferreira for spreading the word around the globe and others who have, on their own initiative, given their support by reviewing this work in the media and on the internet beyond all expectation, Professor Richard Flores, Stan Brown, Dr. Lorna Bernaldo, Dr. Nasser Khalili for his legal guidance, and the Khalili and Outram families for their loving support.

Some of the work of the many students and apprentices is shown in this book and in Cal-Earth's films. Since we have not been able to distinguish every individual by name, we would like to thank them collectively for their contribution and perseverance in building structures all over the world.

CHAPTER 1
SANDBAG SHELTER

Natural Disasters are Human Made Disasters Blamed on Nature.

After a fire, hurricane, flood, or earthquake we immediately call it an "act of God" or Nature's disaster. Then we ask if we have insurance, or how soon will the government or U.N. come to help? And these are repeated and echoed in the media around the globe over and over again. But are these the right questions to ask? Shouldn't we be asking, "Why did our house burn, fall apart, or get swept away?", and when we have the chance to rebuild it why should we build it in the same way and in the same place?

   The human impact on nature and its effects: pollution, deforestation, land mismanagement, the greenhouse effect, global warming and more, will undoubtedly accelerate the rate of disasters in the future. Added to that are man-made disasters: millions of displaced humans, wars and human aggression and terror with its incalculable damage to human life and property which must awaken us to a new set of questions. It is a dire necessity to educate ourselves and our children to act more in harmony with nature, rather than insisting on dominating and interrupting the environmental process.

   To build a simple emergency shelter that will give us maximum safety with minimum environmental impact, we must choose natural materials and, like nature itself, build with minimum materials to create maximum space, like a beehive or a seashell. The strongest structures in nature which work in tune with gravity, friction, minimum exposure and maximum compression, are arches, domes, vaults, apses and organic free forms. And they can be easily learned and utilize the most available material on earth: EARTH.

*"in this earth
in this earth
in this immaculate field
we shall not plant any seeds
except for compassion
except for love"*

- Rumi

The Vision: A Sustainable Solution to Human Shelter

There is a sustainable solution for the accelerating problem of millions of displaced and homeless humans. To create safe shelter we need not cut trees, weld steel or manufacture concrete. In most cases the earth alone will suffice. There is no way that all the world's people will have suitable shelter even if all the industries and governments begin to supply housing. The only way is to teach people how to use the earth under their own feet. Such simple technology based on eco-friendly and indigenous ways exists today.

Superadobe: Sandbag and Barbed Wire Technology

> Who can build with Superadobe?
>
> The whole family should be able to build together, men and women, from grandma to the youngest child. It took many years of researching hands-on how to make the process simpler and easier; no heavy lifting, no expensive equipment, but fast and flexible construction.

What is Superadobe? Superadobe is a super long sandbag, a super long adobe. It can be from standard sandbags or the fabric tubing before it is cut into small bags. It is an earth coil, an instant and flexible line generator.

Superadobe uses the materials of war, sandbags and barbed wire, for peaceful ends, integrating traditional earth architecture with contemporary global safety requirements. Imagine it as long or short bags filled with the earth that is on the site, and arranged in layers or long coils (compression, low-tech). Strands of barbed wire between them act as both mortar and reinforcement (tension, hi-tech), like velcro sticking the layers together. It can be filled with almost any material ranging from pure earth (adobe) to pure concrete, including stabilized earth, and recycled materials. Stabilizers such as cement, lime, or asphalt emulsion may be added.

It can build temporary shelter or permanent structures depending on the type of earth and the climate. It can create both structures and infrastructure from buildings to landscaping such as retaining walls, watercourses and lakes, to cityscaping such as roads, sidewalks, planters and seating. Superadobe can be coiled into vaults and domes,

the way a potter coils clay to make a pot. Barbed wire reinforcement makes the structures resistant to windstorms, floods and earthquakes. Life size model structures successfully passed strict California earthquake code tests.

When we build with arches, domes, vaults, and apses and combine the natural compressive strength of earth with the barbed wire as a tension and friction element between the bags, we can build a whole town from the earth alone.

Superadobe is an adobe that is stretched from history into the new millennium. "It is like an umbilical cord connecting the traditional with the future adobe world."

"Superadobe is an adobe that is stretched from history into the new millennium."

Where Did This Idea Come From?

This concept was originally presented to NASA for building habitats on the moon and Mars with "Velcro-adobe", and later was put into research and development in finding an answer to human shelter here on earth. It comes from an architect rooted in the third world, educated and living in the West. And it comes from the concerned heart and mind of someone who did not want to be bound to any one system of construction and looked for only one answer in human shelter, to simplify, simplify, simplify.

This patented technology is offered free to the needy of the world, and licensed for commercial use.

Right: A child stacks his donut toy rings (concentric dircles).

Shelter Concepts

To build a sandbag shelter, the principle of construction with long or short sandbags is similar to the way a child puts his donut-like toy rings on top of each other. This geometry is one of the most appropriate for Superadobe construction since we can lay the earth-filled coils flat on top of each other, and step them gradually inwards. This is called "corbelling". Equal steps will give us a conical shape, but to create a more spacious interior we use an egg-shaped dome.

In the following pages you will see step by step how to build a small domed shelter or a village using simple earth-bags or long Superadobe tubes, barbed wire and earth. You will learn how to:
1) Use the materials of war (sandbags and barbed wire) for peaceful purposes by creating natural safe shelter and infrastructure.
2) Utilize a minimum amount of purchased product and maximum amounts of nature's gift of the earth under your feet, to create a low cost shelter and protect against natural or man-made disasters.
3) Participate in a family or community activity by building a shelter or village.
4) Be empowered to build and help others without any previous construction experience, and be confident of sheltering yourself and your family in any emergency.
5) Go anywhere in the world and use the natural elements you find around you to build.

Building Permits:

In most counties in the U.S. with enforced building codes, a homeowner can build an accessory structure of about a hundred square feet, without needing a building permit; you may not be allowed electrical or plumbing. Check your local codes. Most of the shelters shown in this chapter are less than this size.

Certain hands-on exercises will prepare you to build a domed shelter:
a) Building an arch from sandbags you will learn the structural principles of all domes, vaults, and apses.
b) Filling rows of long or short bags with earth/stabilized earth and placing 4-point barbed wire between them, you will learn the principal materials and technique. The barbed wire acts like mortar or velcro sticking the bags together.

Start with a small landscaping project if you feel you need to get more experience building arches and low-rise walls. You should feel ready to start your first shelter when you have practiced enough by filling bags and laying barbed wire.

The following single shelter was built by students with no previous experience, who were learning for the first time. Yet despite this, it is a safe shelter. More refined techniques may be learned after gaining this experience, which are shown in the Eco-Village shelter variations.

This manual is used with the Cal-Earth hands-on apprenticeship training course, but for those who may build with this book alone, we ask you to use your own judgement and common sense. Although Superadobe is a simple technique, you must make the first building small enough to learn safely. Respect gravity and don't take any unneccessary risks. Learn from the local tradition, since local conditions of earth, climate and building techniques vary. Be responsible for the safety of your team and yourself as you learn.

Practice, practice, practice, until you get it right.

Above and below: The emergency shelter dome concept. Coils of earth-filled bags stepped in a little at a time to form an egg-shaped dome.

For an immediate, quick learning guide, without reading the rest of this book, the following four pages, which shall be kept updated, are a summary that can be copied onto the front and back of one sheet of paper, strong enough to resist construction site damages.

This sheet can also be laminated in plastic.

Translation into local languages and sending a copy to Cal-Earth is encouraged.

*"every man and woman
is a doctor and a builder
to heal and shelter themselves"*

*Superadobe
sandbags
earth-bags
earth coils
are all the
same thing.*

EMERGENCY SANDBAG SHELTER: QUICK TRAINING GUIDE

10 STEP SUMMARY FOR BUILDING A SINGLE DOME

1. Bring your TOOLS & MATERIALS - bags, barbed wire, etc.

2. Check what kind of EARTH you have and prepare for use (stabilized or unstabilized).

3. Decide on the DOME LAYOUT based on landscape and climate.

4. FOUNDATION: Draw a circle (dome plan) with a center compass, dig a foundation trench and fill with the foundation bags. Learn how to fill and tamp bags, and place barbed wire.

5. BASE: Fill bags for a short stem wall. Leave a door opening and start the door walls for the entry.

6. DOME: Set up the compass to build the dome. Build the dome by coiling the bags all the way to the top using a compass (two chains) to measure the shape.

7. DOORWAY: Set up a door frame or guide for door the opening. Continue the dome bags up to the frame and then over the top of the frame as a small lintel or arch, and support it with barbed wires (and branches or pipes if needed).

8. WINDOWS: Arched windows - make a form with loosely-filled bags or other materials, build the dome around the form and remove the form when the dome is completed. Make an "eyebrow" over larger window openings. Pipe Windows - build the dome over the pipes for ventilation and light.

9. BUTTRESS WALLS: Build the buttress walls for the entry and connect to the dome at every row with barbed wire. Build as steps for climbing to the top of the dome.

10. ENTRY VAULT: After the upper dome is completed, use leaning arches to complete a small entry vault to protect the doorway. Or use very short lintels for a "mineshaft" entry.

EMERGENCY SANDBAG SHELTER: QUICK TRAINING GUIDE

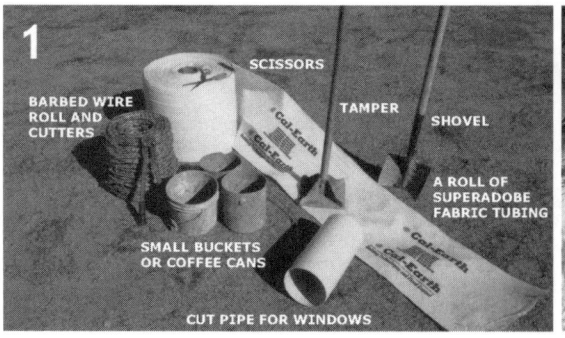

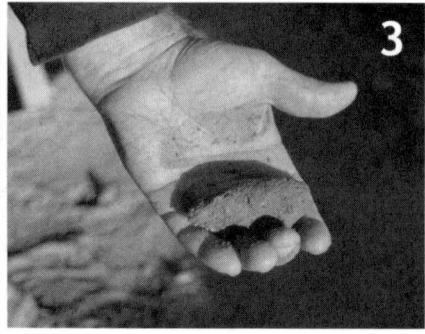

1) Collect the tools above. 2) Prepare the earth mix which is stabilized with cement, lime, or asphalt emulsion. 3) Add enough water to ball together when squeezed, yet not leave the hand wet. If no cement, lime or barbed wire is available, use raw earth (clay-sand) for a temporary shelter. (Experimental: try snow in bags and compact.)

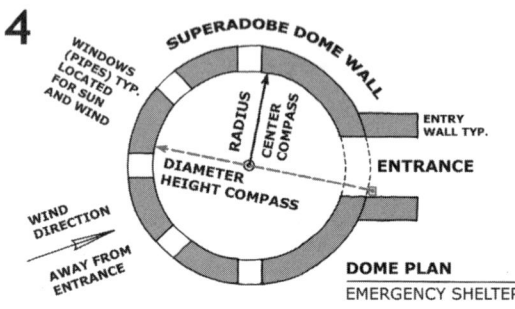

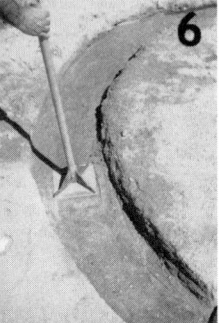

4) Place the door away from wind and rain. 5) Dig the foundation trench 12" (30 cm) deep. 6) Level and compact the trench. 7) Place the bag in the trench, fold the end under to close, and start filling upright like a short column (using long tubing or standard sandbags). The foundation will be 2-3 completed bag rows.

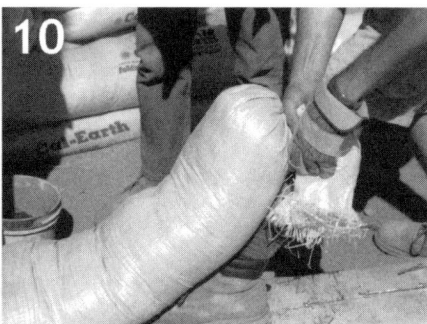

8) Always put in 2-3 cans of earth and shake to the end. 9) Use gravity's help by sloping the bag on your leg, and walking backwards as it fills. Do not strain (if it is hard and feels heavy you are not doing it right). Let the bag fill as full as possible and check the position with the compass tool. 10) Twist and tuck under the bag ends to close.

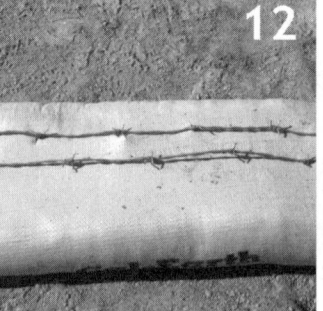

11) Compact the filled bag as hard as you can using a tamper, to make a smooth, solid, uniform block. Only compacted damp earth becomes strong. 12) Attach continuous barbed wire: 1 wire for domes up to 12 ft (3.7m), 2 wires for larger domes. Where breaks occur, overlap the wires by 2ft. (65 cm). 13) Continue coiling bags.

Architect Nader Khalili

Cal-Earth Institute - www.calearth.org

EMERGENCY SANDBAG SHELTER: QUICK TRAINING GUIDE

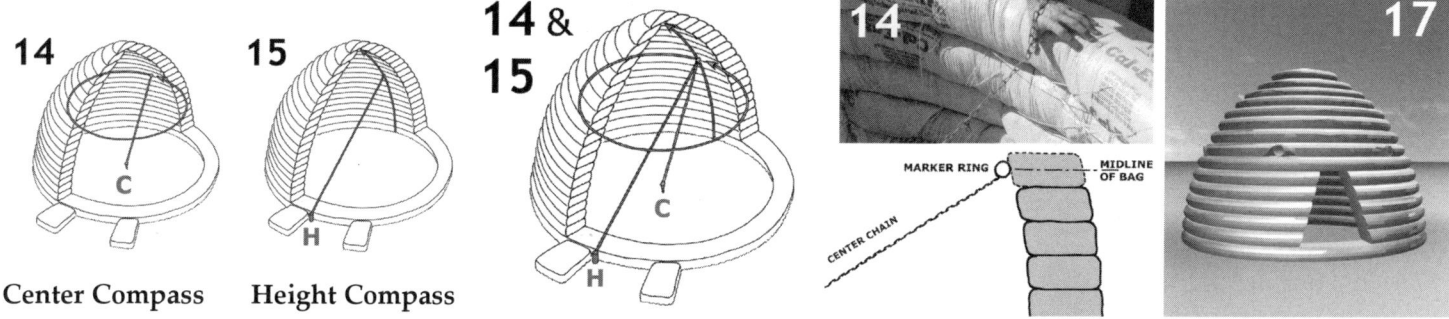

14, 15) You must use 2 Compasses to make the dome shape. Use chains or other non-stretchy cable-rope. Attach one in the center (Center Compass), and extend the length at every row according to a second one at the perimeter (Height Compass). <u>If any bags do not conform to the compass remove them and re-build.</u>

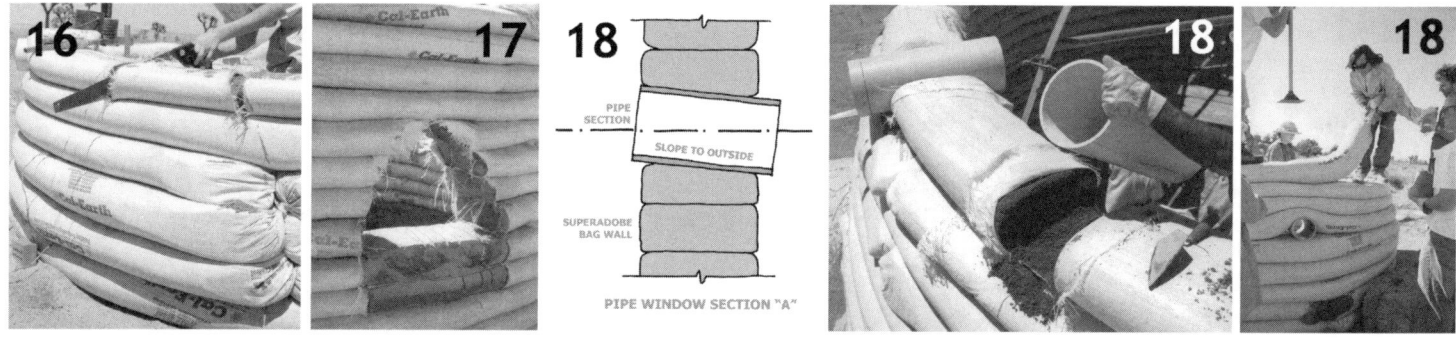

16) Pre-cut bags for a doorway knock-out panel, or for windows. If stabilized, the earth must be cut after tamping at every row; don't let the cut bags stick together again. 17) Punch out pre-cut panels to open the door or window after at least 5 bag rows above are completed. 18) Insert pipes for vents or windows, sloped to outside for rain.

19) Coil upper rows, but don't stand on the wet bag. 20) Fill and place each bag parallel to the row below and work it cautiously inwards to meet the compass circle (pull or tamp). Tamp the bag with a gentle slope to the outside. Close the top or leave a small skylight. 21) Add a protective entrance (door vault) to buttress the door opening. Entry is arched or sloped, short or tall.

22) Plaster the exterior before bags disintegrate and 23) waterproof with locally suitable materials to resist moisture and erosion. 24) On top, finish with a water-resistant cement or lime plaster such as Reptile (stabilized earth mud-balls) layered from bottom to top, or 25) a smooth cement or lime plaster finish.

Architect Nader Khalili   Cal-Earth Institute - www.calearth.org

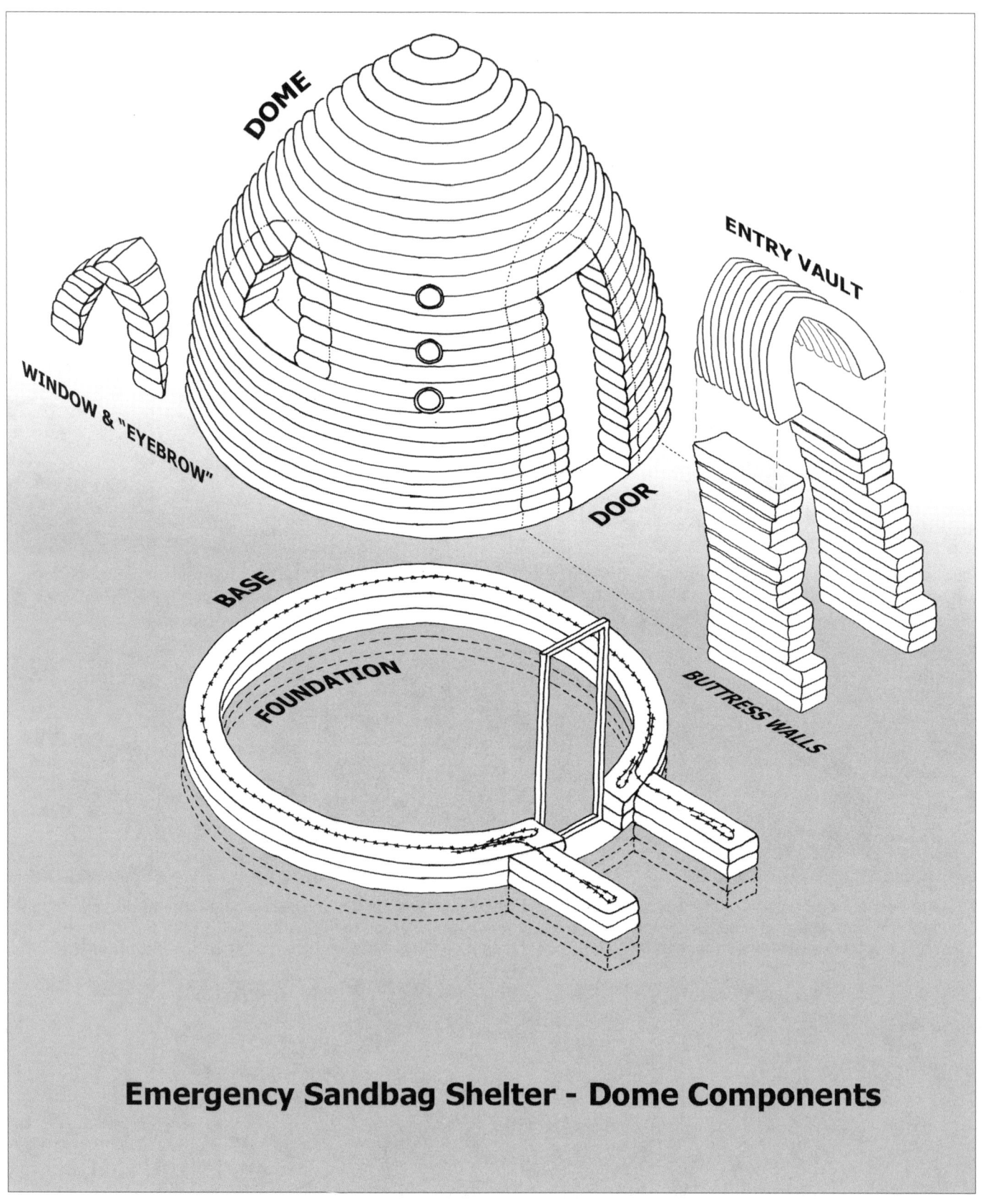

Emergency Sandbag Shelter - Dome Components

*"every man and woman
is a doctor and a builder
to heal and shelter themselves"*

A SINGLE DOME

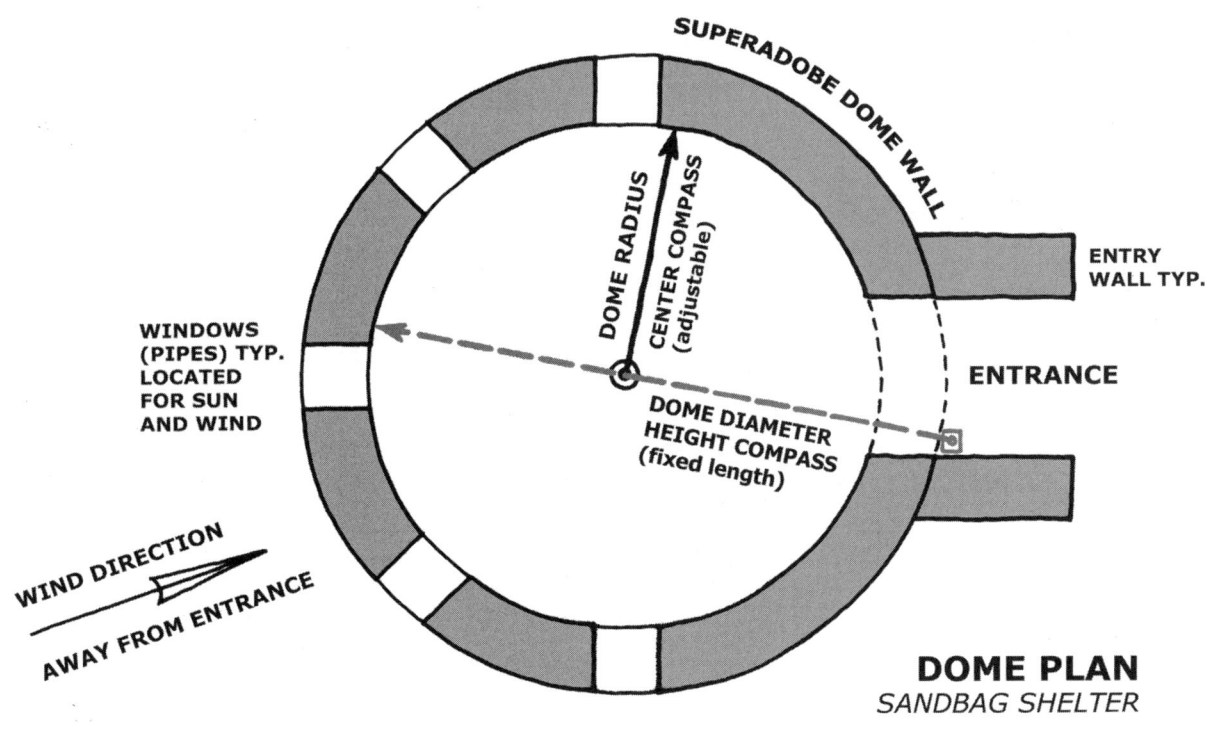

DOME PLAN
SANDBAG SHELTER

Above: When earth is piled up it naturally forms a cone, the shape of a mountain. The conical or domed shape is the natural structure of earthen material in harmony with gravity. And its slope is called the earth's "angle of repose." This is true within the gravity of planet earth, and also true on the moon or Mars and beyond, with their own angle of repose.

It is a privilege to dig the earth from under our feet and build. A pile of earth becomes a playground for the designer's imagination, from which a shelter will be created for the body and soul.

Right: Digging the earth under your feet, and filling small pots or buckets (one small pot equals about one shovel measure).

Above: A shelter is hand built by students, digging and passing pots of earth up to the builders.

From a pile of earth a shelter grows, with human knowledge, healthy labor and team work. Shelters can be hand built, one small pot at a time, or mechanized using pumps, one coil after another.

Left: The pile of earth diminishes, while the domed Superadobe/sandbag shelter grows from the earth.

SUPERADOBE - SANDBAG SHELTER

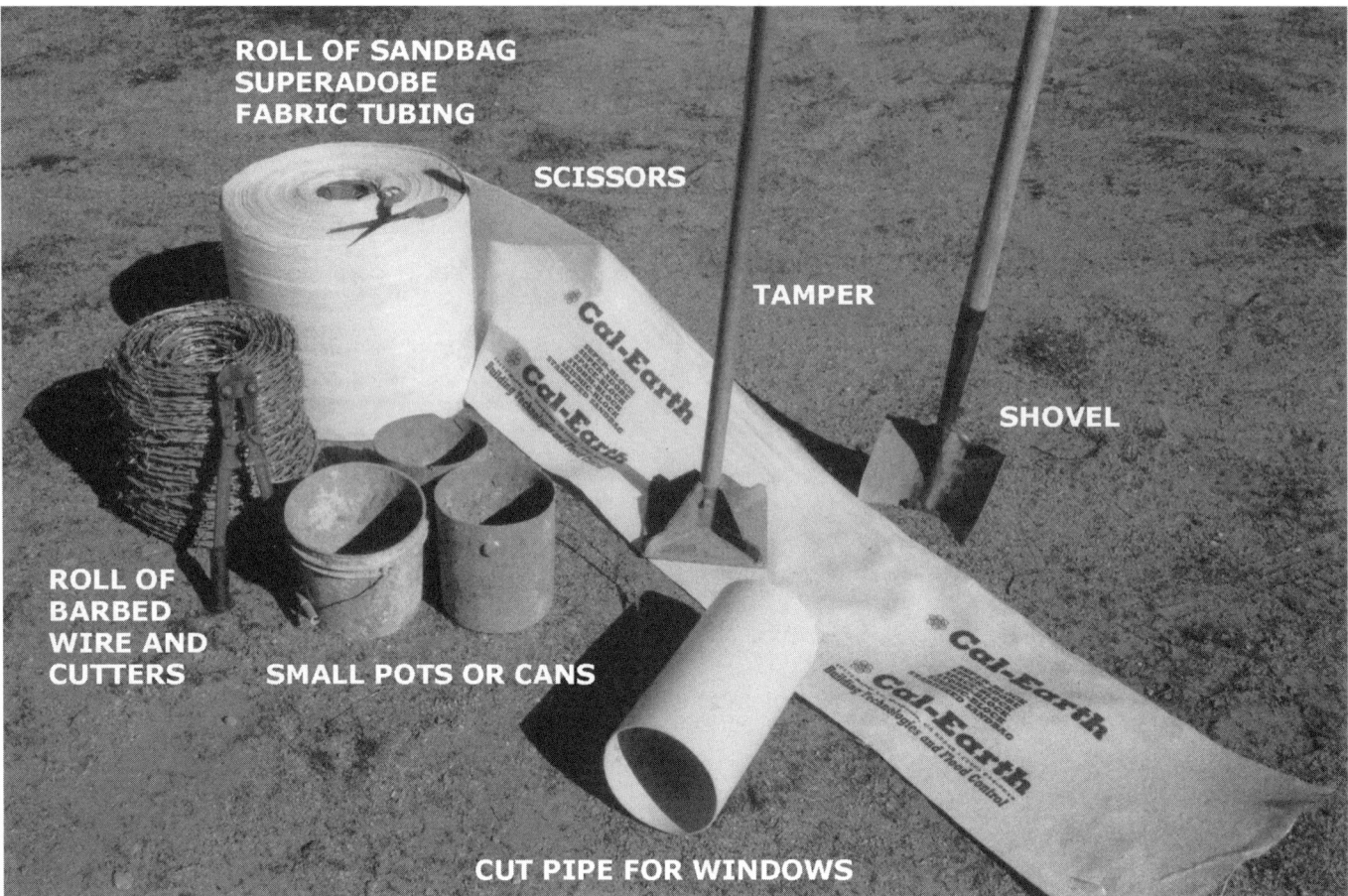

Above: The tools needed to build.

Long Bags (Tubing)

The long bags which make Superadobe coils are the fabric tubing before it is cut and sewn into small standard sandbags. The material most commonly used today for sandbags is woven polypropylene, a type of plastic material. This material, depending on its UV (ultraviolet) rating, will disintegrate when exposed to direct sunshine. Commonly used bags have an average of 300 UV, which means that they will start to break down after 300 hours of exposure to sun.

Natural fiber bags, such as burlap or jute fabric are normally sprayed with toxic chemical preservatives such as formaldehyde. They are also more prone to mold and insects.

To safekeep the polypropylene bags you must first store them protected against sunlight. After construction and to last a long time, even years, you must plaster any surface which is exposed to sunlight before the 300 hours is reached. The polypropylene fibers making the bag are extremely strong when new. You should not be able to break three or more strands together no matter how strong you are.

While this book teaches how to build with the long bags, the same structures and landscaping can be built using small standard sandbags (see Appendix III: Small Bags).

Above: Set up your bag roll and barbed wire to easily unroll and cut off measured lengths.

Tools and Materials: Preparing to Build

Gather your tools and set up your construction area:
1) A roll of 14-16 inch wide Superadobe tubing, or small sandbags. (The labels on the tubing are 3 feet apart, or one pace, to guide with measurements.)
2) A roll of 4-point, two-strand barbed wire.
3) A compass of 2 thin chains longer than the dome diameter, six metal rings, and two posts or stakes (see Appendix IV: Compass).
4) Shovels.
5) A knife or scissors to cut the bag material (scissors preferred).
6) Pliers or bolt cutters to cut barbed wire.
7) Small empty pots (coffee cans or small buckets are a good size).
8) A tamping tool (plumber's tamper or home made tamper).
9) Several 6 - 8 inch diameter pipes, about 14 - 20 inches long.

The following additional elements are optional:
10) Portland cement or lime cement (for stabilized earth).
11) Work gloves (plastic gloves for stabilized earth and reptiles).
12) Plastic sheeting or a tarp.
13) Waterproofing compound such as asphalt.
14) Trowel.
15) A saw or metal wire.
16) An arch form.
17) Glass or plastic for windows.
18) A door frame.
19) A door panel of wood or metal; a carpet or heavy cloth.
20) Any other tools shown in this book are also optional.

Below: Wooden temporary arch forms of different sizes. A good form has a handle (a block or hole).

SUPERADOBE - SANDBAG SHELTER

Right: Dry earth on the site.

Below: Digging down for a supply of earth. After rain it is moist and ready to use.

Earth/Soil Samples and Tests

To find out what kind of earth (soil) you have on your site you need to do some simple field tests. Mixing the earth in a jar of water and letting it settle will show you its composition in layers. At the bottom will be gravel and large grains of sand, followed by smaller grains of sand, then silt and finally clay at the top. The clay which suspends in the muddy water may take several hours to settle.

Based on this jar test, you must decide whether you are going to:
a) Use moistened earth alone for a temporary structure (if it has enough clay to stick together well).
b) Stabilize your earth with cement, lime, asphalt emulsion, or other stabilizer, for a more permanent structure.

For permanent structures such as these emergency shelters of 8-12 ft. diameter, the earth mix should resist erosion by water and

Right: Stabilized earth samples in plastic cups.

Below: The jar test. Place a handful of earth into a jar of water, close the jar, then shake to mix well.

Left: After the earth has settled, look for layers from gravel and coarse sand, to finer silt and clay.

Note: For mass emergency shelter projects, a professional soils laboratory should determine the right mixture.

should reach at least 300 psi (pounds per square inch) compressive strength, similar to traditional adobe bricks. Samples must be made to test the mixture's strength and water resistance.

The following steps determine a suitable stabilized earth mixture.
1) Choose your stabilizer. For sandy soil, cement or lime are best. For adobe/clay soils, lime or asphalt emulsion are more suitable.
2) Make a sample mixture. For example, mix 10 cups of earth with 1 cup of cement and add water to a moist, muddy consistency.
3) Pack this mixture firmly into three plastic cups.
4) Let the samples dry in the shade.
5) After several days, take off the cups and put the samples into a bucket of water.
6) If the samples do not erode in the water after 3 days, then that mix is good for filling the bags.
7) If needed, reduce or increase the percentage of stabilizer.

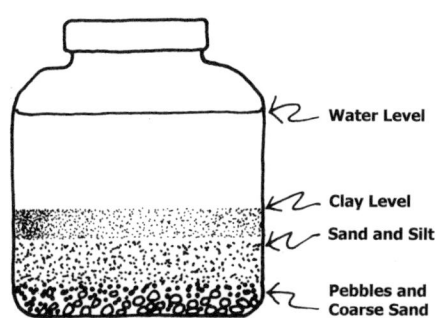

SOIL TEST JAR

Left: Testing a stabilized earth sample in water.

SUPERADOBE - SANDBAG SHELTER

Right: Adding water to earth (no cement).

Above: Measuring a can/bucket of cement or lime for making stabilized earth.

Right: Mechanical mixing of stabilized earth in a plaster/concrete mixer.

Using the Earth Mix

Stabilized earth may be mixed by hand or by machine. First, remove any large rocks but keep the gravel.

The cement or lime must be consistently mixed with the earth and water. Portland cement is preferably mixed dry with the earth before adding water. Lime must be first mixed with water, then mixed with the earth. Before using the stabilized earth, check that the stabilizer is thoroughly mixed with the earth and evenly moistened.

For unstabilized earth simply add water until it is evenly moistened.

Left: Mixing earth, cement, and water in a wheelbarrow to make stabilized earth.

Checking the Mixture

The earth mix should be moist but not wet, to ensure the maximum density of the earth particles. When you squeeze the mix in your hand it must form a ball that does not fall apart yet does not drip water.

If it is too dry, the earth will not compact when tamped, and will not harden inside the bags. If it is too wet it becomes fluid when tamped, making it difficult to build and weaker when it has hardened.

To assure yourself that the stabilized mixture is suitable for building, squeeze several samples in your hand.

Above: Stabilized earth mix ready to be used.

Left: Squeezing to check if the earth is ready to be put into the bags.

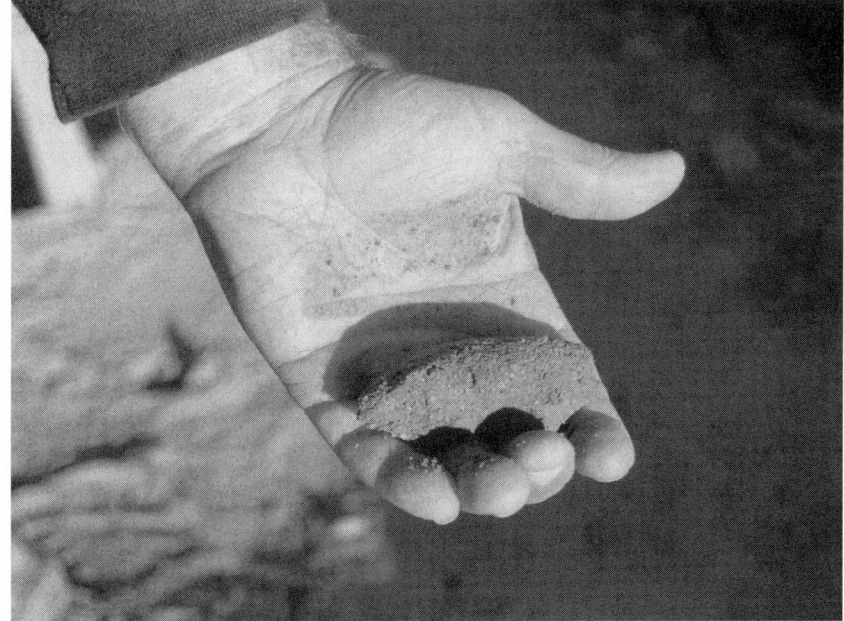

SUPERADOBE - SANDBAG SHELTER   31

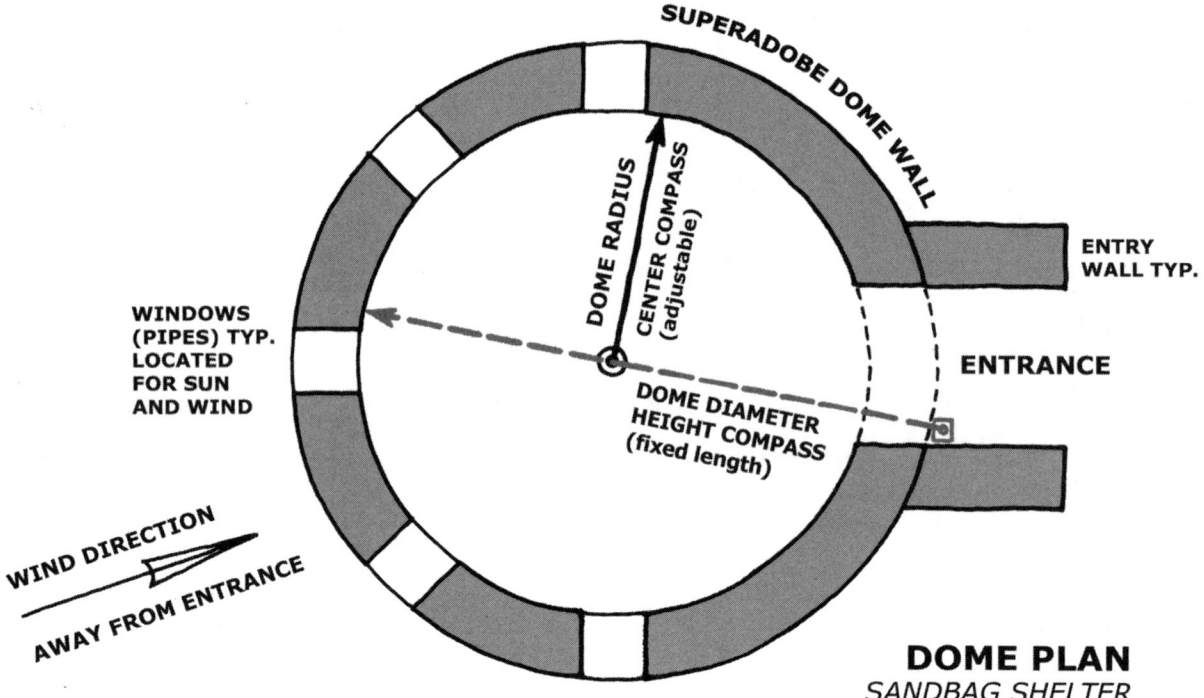

Above: Plan and orientation of a typical emergency shelter dome.

Note:
1) See Appendix IV: Compass.

2) The entry walls can be added at the end of the dome construction, or they can be built along with the dome for a more permanent structure.

A SINGLE DOME: Layout and Orientation

An 8 ft. to 12 ft. interior diameter (2.5 m to 3.5 m) dome uses 14" inch to 16" inch wide bags and can be built in one to seven days by a team of three to seven, one of whom is trained in the system.

To locate a dome/shelter on a piece of land, a few basic principles must be considered:
  a) Place the door away from harsh winter winds, but let your windows catch the summer breeze.
  b) Locate the door and windows away from the hot summer sun from the west, yet invite the warming sun from the south in winter (northern hemisphere). Find shade for hot climates.
  c) Place the dome away from water such as streams, swampy ground, or rainwater runoff, and avoid soft organic soils.
  d) Building on a hillside, you must provide enough footings and retaining walls (including weepholes) to prevent the dome from sliding down the hill or being exposed to mud slides.

Clear and level the building site

Clear away any plants and organic materials such as compost, leaves, twigs, or mulch. Organic materials should not be put inside the bags nor should foundations be built on it. Dig down until you reach the natural earth and set aside all the organic material.

Left: Clearing and levelling the site. Three people are an ideal building team, however it is possible to build a dome with two or even one person.

When the area is cleared you can level the earth ready for the base of your dome.

Draw a Circle

Geometry in these structures starts with one circle. To begin the emergency shelter, you can take a piece of rope, chain, or cable, and two short sticks, and start by drawing a circle on the ground. Often we ask children and adults to lie on the ground, spread out their arms to measure their body's length, then add two more feet. This will be the size of their shelter.

If your shelter is eight to ten feet in diameter, you must set the length of your chain to four to five feet and use it as a radius to draw the circle.

Left: Drawing a circle on the earth using a center stake, a chain, and a rod.

Above: Hammering the center compass stake into the ground.

Right (above): Marking two circle lines on the ground, to position the dome wall before digging the foundation.

Set up the Center Compass

A compass is needed to create the geometry of the dome. This is a tool attached in the dome center which measures every ring of Superadobe as you build.

In this chapter you will see a) a chain compass, or b) a piece of pipe which swivels around the dome center. See the "Compass" chapter on how to set up your compass.

In the beginning, the center compass is marked at the dome radius length.

Using the center compass, mark two circles on the ground for the inside and outside of your Superadobe dome wall.

Right: A chain is attached with a ring onto a metal center post. Two metal links allow it to turn freely.

Above: Levelling the trench with a carpenter's level.

Left (above): Digging the foundation trench.

Foundation: Dig a Trench and Level the Base

Dig out between the two circles to make a foundation trench around 12 inches wide and 10 inches deep (for an 8 - 10 ft. inside diameter dome). Put the earth to one side for building later, and make sure that the trench is level and well compacted. You can check the level using a carpenters level and a straight length of wood, or you can use a water level in a length of clear tubing, since water always finds the same level in gravity. By partly filling your foundation trench with water, you can quickly see if any areas are too high or too low.

A small shelter up to 10 ft. inside diameter, with thick walls, will usually not need a special foundation. However, it is wise to avoid expansive clay soils, frozen subsoil, soils which are eroding and subsiding, or very wet subsoils. For larger, more permanent structures the foundation design must be adapted to cope with the particular site conditions.

Left (below): Using water to find the level.

Below: An alternative to the chain, a pipe compass arm is attached to a post and turns freely.

SUPERADOBE - SANDBAG SHELTER

Above and right: Tamping the foundation trench after it has been levelled and the water has subsided.

Compact the Base

The base of the foundation trench must be moistened and firmly compacted. The tool being used is a plumbers' tamper which is generally used to compact earth into plumbing trenches. If one is not availble, a similar tool can be made for tamping. Overall your trench should be fairly level, but minor undulations are acceptable.

A deeper foundation trench, for building partly into the earth, improves the structural stability of the dome and gives better insulation from heat and cold. A range of 6 -18 inches depth is ideal for an 8-12 ft. diameter dome. However, rain water must be diverted away from the entry doorway, making deeper foundations less practical.

The trench should also be dug for the entry buttress walls, which should be built along with the dome. (In a temporary dome, a small entry with walls can be added after the dome is completed, for a quick emergency method.)

When building into very dry earth the trench should be soaked with water during construction, to allow stabilized earth to cure.

Right: A tamped and finished foundation trench, one to two bags into the ground. The foundation trench for the dome and door vault are dug together.

First Superadobe Rows

The foundation is two complete rings of sandbag coils with a layer of barbed wire in between. We need two complete rings to stabilize the foundation of the dome, especially for horizontal forces like earthquake, windstorm and flood.

Fill the Bag

First, cut off a length of sandbag tubing from the roll using scissors (a knife may shred the bag). You can pin the end closed with a nail, or you can tuck it under and let the bag's own weight hold it closed.

Next, fill the bag with your damp or stabilized earth mixture. There are several methods for filling the bag. Here it is being filled at one end with small cans of earth, about two or three at a time. You can take the open end to the pile of earth, or take the earth material over to the tube. Your team can continuously feed the bag and shake it to the building spot, like a swan or snake swallowing bite after bite.

Above and left: Unrolling and cutting a length of Superadobe tubing

Below: Digging up the moistened earth and putting into small cans/buckets.

Left: The mouth of the bag can be taken to the source of earth and filled if several people are participating.

SUPERADOBE - SANDBAG SHELTER

Above: The end of the bag can be pinned closed using a nail, or simply folded and tucked under.

Right: Filling the bag.

Gravity Flow

At first, the bag is filled vertically like a short column so that gravity helps to fill the start of the bag very full. This will compensate for the tendency of bags to get too thin at the end of the row.

Below: Filling the mouth of the bag with cans of earth/stabilized earth.

Right: Filling the bags from two sides of the dome to move faster.

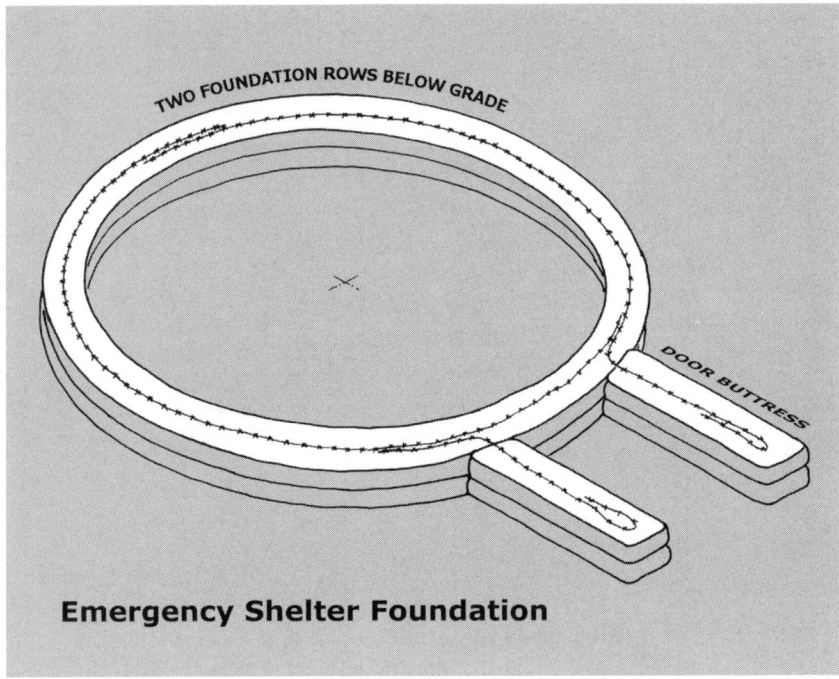

Emergency Shelter Foundation

Above: The center compass measures the inner edge of the bag as it is placed in a circle.

Left: Plan (axonometric) of the two foundation rows below grade/ floor level, showing the barbed wire layout.

The builder then continues to fill the bag with gravity's help, sloped up against his leg or other prop, while walking backwards.

Check the Circle

While the bag is being filled and placed by the builder, it is constantly checked with the compass which measures the distance from the center to make a circle. At this stage the inside edge of the bag will be about half an inch or one finger width beyond the compass line.

Below: Before taking a break, the student packs the earth with a squeeze and sets down the bag. Note the center compass here is a piece of pipe, or a stick.

Left: A new student learns to fill the long bag at a slope while walking backwards.

Above: Tamping the bag after it is filled and checked with the compass.

Tamp the Bag

Tamping means compacting the earth inside the bag very hard to make a uniform ring. This gives the earth mixture its strength.

After the bag is completely filled, it is tamped using a plumber's tamper or other tool. But first, before tamping the bag is adjusted one last time to make sure that it fits the compass curve leaving one half inch gap to allow for spreading during tamping.

When compacting the bag, you will see that it spreads or widens and also flattens. A bag of 14 to 16 inches wide is generally used for a small shelter of 8 -10 ft. diameter. After filling and tamping it will become about 12 - 13 inches wide and 5 inches high, and should fit the compass exactly. To check if your bag is sufficiently compacted, press your finger into the outer edge. If it dimples, tamp some more.

Tamping may seem like hard work in the beginning, but is a fun exercise. To tamp without getting tired or hurting your back, loosen your grip on the handle just before the tamper hits the bag. The force of the tamper itself will compact the earth material, and your hands will merely steady the handle.

Left: Twisting to close the filled bag.

A brick may be used to compact smaller detailed areas, for the sides of the bags, or for the whole dome if necessary. In an emergency, human feet can tamp sufficiently for small temporary shelters using unstabilized earth.

Good and Bad Tamping

To check if your tamping is good enough, look at the outer edge of the bag. If it is a smooth curve it is good tamping. It should feel firm. If it is soft, the bag needs more tamping.

When the earth mixture is not tamped enough the sides and ends within the bags will crumble and are weak. Stabilized earth will not stick together properly if it is not well tamped, since the earth mixture used for Superadobe is damp but not wet like concrete.

Above: Tamping the closed bag row.

Left: The completed foundation row. The extra bag may be used for the next row.

SUPERADOBE - SANDBAG SHELTER

Right: A roll of 4-point, 2-strand barbed wire will build about two small domes.

Barbed Wire

After each row is completed, a strand of barbed wire will be placed over that row.

The barbed wire is tensile reinforcement for the earth wall of the dome. This means that the barbed wire resists tension forces which exist in the dome. These tensile forces are a result of the static forces created by it's own shape, and of dynamic forces which result from liveloads and seismic forces. If there are breaks in the continuity of the wire, it is less effective. A small dome of up to ten feet or three meters diameter needs only one strand of barbed wire and will be strong enough with either thicker or thinner gauge wire. However, any larger buildings need two or more strands.

The barbed wire must be four-point, two strand, and galvanized. If there are only two points on each barb the wire will not grip well enough. The four-point barbs on the wire will increase friction between the bags to stop them slipping over each other from horizontal shear forces. The double strand of barbed wire will take up tension in the superadobe wall to resist diagonal cracks forming in the wall and buckling outwards.

Below: Pliers or bolt cutters are used to cut the 4-point, 2-strand barbed wire. Gloves must be used to handle the barbed wire.

Right: Continuous 4-point, 2-strand, galvanized barbed wire grips the bags below and above.

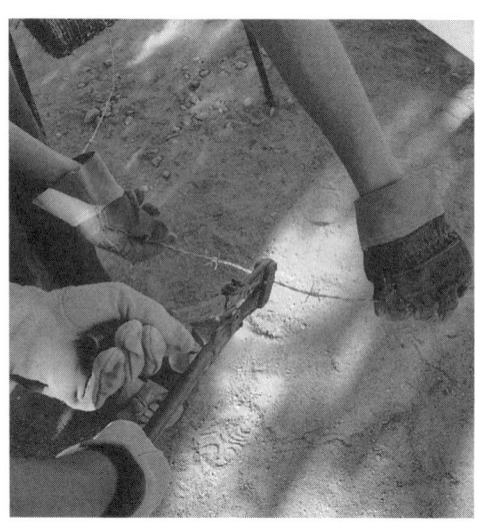

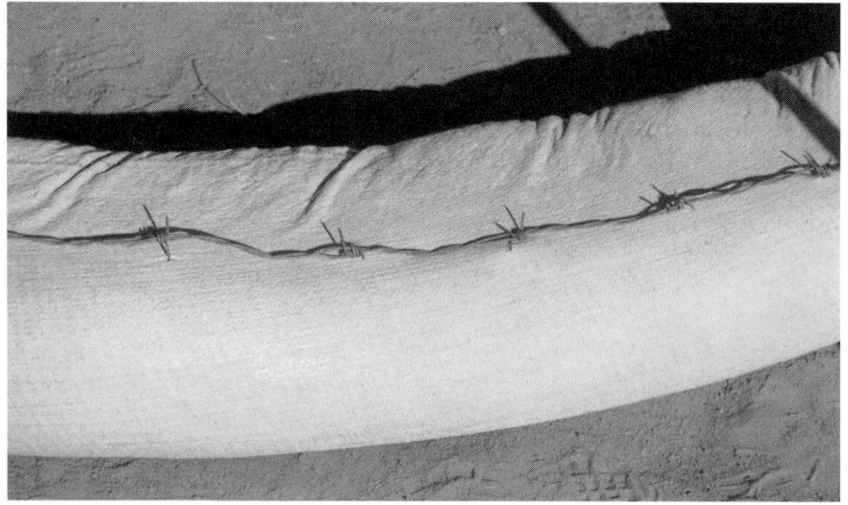

Above: For a small dome a single strand of barbed wire is positioned to allow the next row to cover it completely (by at least 3 inches). The barbs grip the bag.

Below: Bricks can temporarily position the barbed wire.

To use the barbed wire, first unroll it carefully wearing gloves and straighten it out to handle more safely. Cut the length you will need by holding the wire on both sides of the cutters to be sure that it will not recoil dangerously. Sometimes one can continue the barbed wire into the next row without cutting.

The barbed wire is placed in the center of the bag, not less than 3 inches from the edge. To attach it to the bags you can push in the barbs using the wire's own twisting motion to grip the bag and hold itself in place. You can use brick weights, or pieces of wire to fasten to the bag, or even nails, staples, or tape.

One length of wire should go all around the ring without any breaks. Wherever there is a break in the wire, the two ends must overlap by at least two feet, which are twisted together or laid side by side. Twisting together is better.

The second foundation row of the dome is also one completed ring of Superadobe coil which is laid over the barbed wire. Thus a continuous barbed wire tension ring will lie under the doorway sill. Always check the circle with the compass both while you are laying the bags, and after compacting them. Because the bags get flatter and wider after tamping, always think ahead when you measure with the compass. Every row of Superadobe must be compacted well, and between every row there must be a continuous strand of barbed wire.

Generally, the design of your dome must prevent the barbed wire from getting wet. Additionally, when it is sandwiched between inert plastic bags the galvanized steel is less prone to oxidation.

SUPERADOBE - SANDBAG SHELTER

Above: Completed foundation ring.

Below: Barbed wire joins two bag pieces.

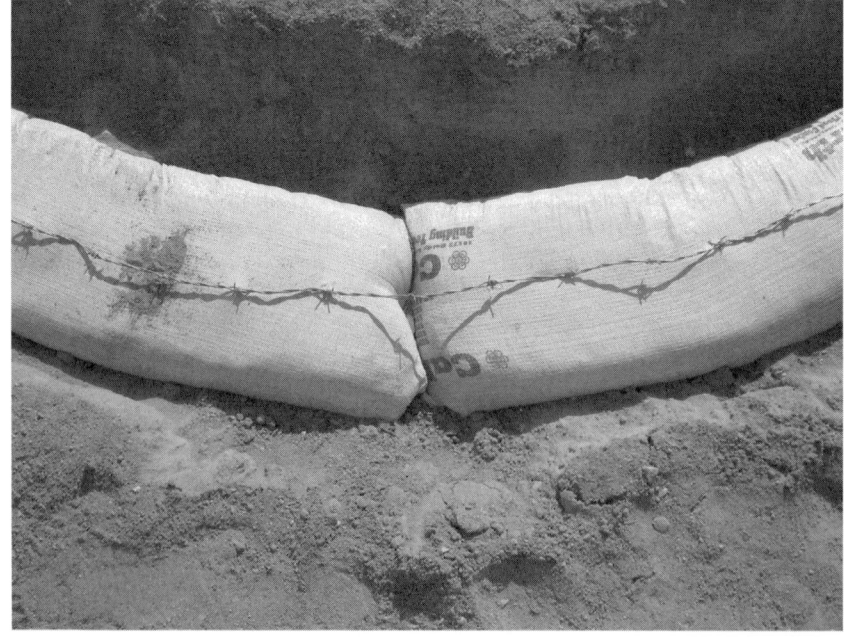

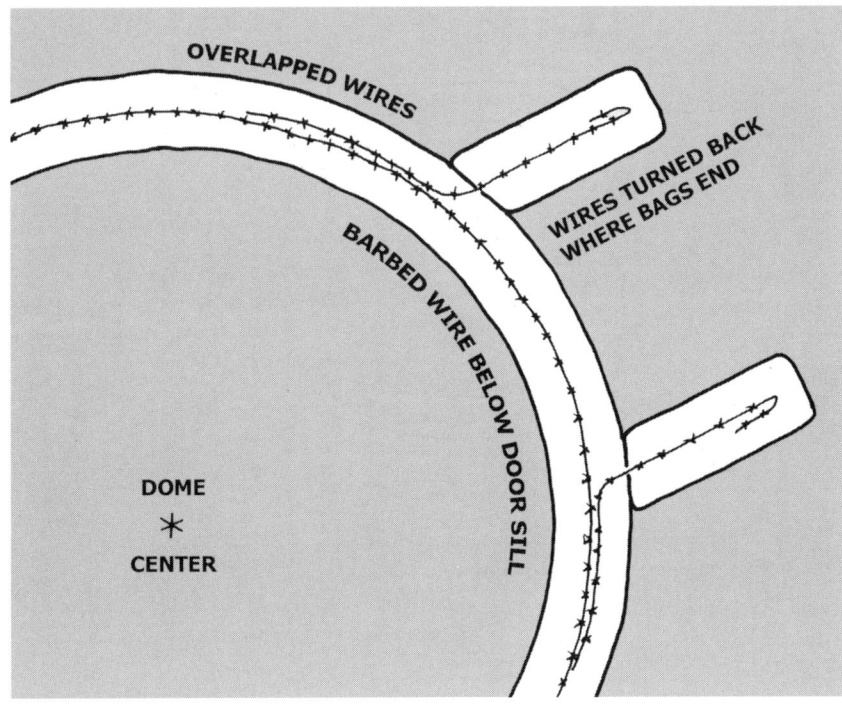

Left: Plan detail showing how to connect buttress walls.

Two short walls for the entry way are also completed for the foundation and tied into the dome with barbed wire.

As you can see from the photos, when two pieces of barbed wire join they overlap by at least two feet and are twisted together. The continuous wire also spans over areas where separate bags join. At the ends of walls, and when we later build the windows, the barbed wire is looped back to give extra friction in these weaker areas.

Left (below): Wires overlap where two walls join.

Below: Wire loops back where bag ends.

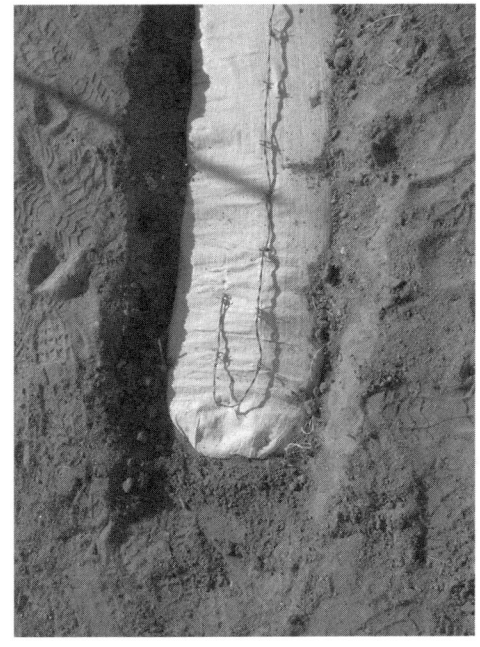

SUPERADOBE - SANDBAG SHELTER

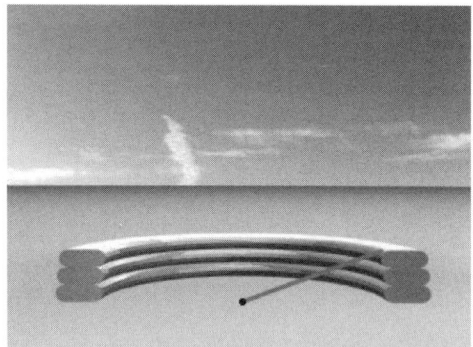

Above: The compass changes very little for the second and third rows of the dome.

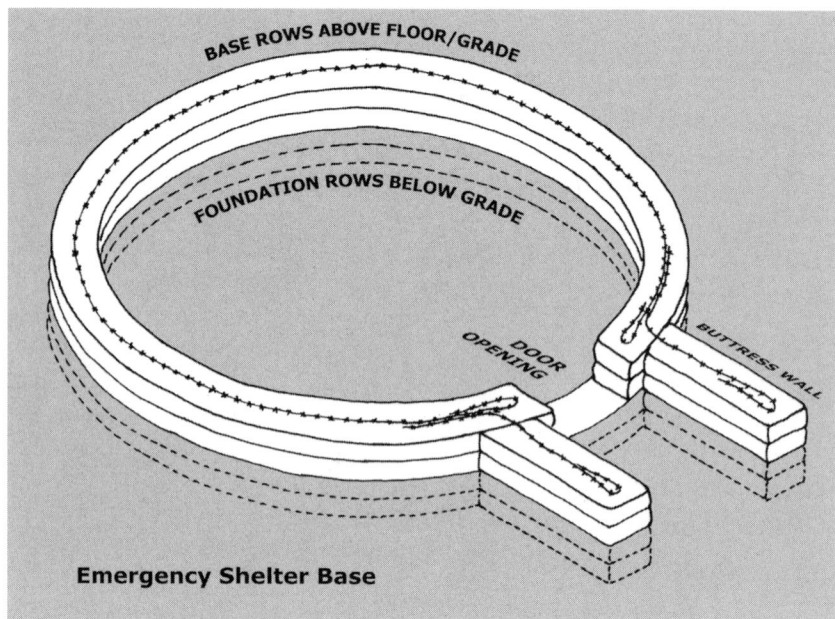

Emergency Shelter Base

Base: Straight Wall Rows

The dome base and foundation together make a low cylinder and using the compass makes sure this is circular. Since the foundation and the base rows sit directly above each other, a plumb bob or carpenter's level can be used to keep them vertical.

After the second foundation row we must leave a space for the door opening (or we can continue the completed rings and make cuts to knock out the door bags later).

For the foundations or walls, not all layers need to be level horizontally and may undulate thicker and thinner depending on the skill of

Below: The builder holds the cut end of the bag tubing closed with his foot when starting to fill the tubing. Later, the weight of the filled bag itself will hold the cut end closed.

Right: Builders in training complete the second row of the base, which is above grade.

Above: Shortening the empty bag length by doubling it inside out.

Left : Supporting the bag with a leg so that gravity helps to fill it.

the builder. To keep an overall level for the whole building, the thickness of the layers may be adjusted row by row because the Superadobe is a flexible block.

More Bag Filling Techniques

As you continue, you will learn several different ways of filling sandbag tubings, and a great many have been tried at different stages of research. Here are a few techniques that have proved useful:

When using long or short bags, always fill the bag in place on the wall. Stand on the wall while working. Use a small can of earth instead of a shovel to fill the bags; it contains the same amount and can be lifted up higher, or even thrown up to the builder. Let gravity help you and don't strain against it or lift weight. Slide the earth into the bag with gravity's help. Save your back and use your leg muscles to support the sloped bag and to shake the earth into place; thus your body will be better balanced. To help the "gulp" go down better, just lean backward and pull the front of the bag fabric material, and also twist the bag to go around the curve. Don't lift the weight of the filled bag. Move the earth to where you want it when it is loose and "flowing" and give it a squeeze to fix it there. Walk backwards as you fill.

The fuller you make the bag, the less work overall, since you will need less rows to complete the dome.

Using the long bag instead of small standard sandbags has already cut your working time in half, and is highly recommended. However, this dome can also be built with small bags. An efficient workflow can keep up the momentum of coiling and building. With a team of three

Below: Pulling the front of the bag material to let earth slide down into the bag.

Right (above): Resting the bag on a leg to fill with gravity's help.

Above: The bag is folded back to shorten the overall length, then unfolded as it is needed.

Right: One long bag has been filled from both ends at the same time and the builders meet as they close the circle.

or more people there can be a constant supply of earth flowing down the long bag for the builder to pack into place. You can organize two or three people to supply the cans of earth, shake the earth into the tube, and pass this earthen "gulp" to the builder.

One long bag has two ends which can be filled at the same time by two different builders.

You can shorten the length of the empty tubing by pulling it inside out to make half the total bag length, or scrunching it up on a tube and passing the free end back through the tube for better control of the bag length as it is filled.

One student got tired of using his leg as a support and innovated a "wooden leg" substitute in the same shape as his own leg.

Left (above): Using the "wooden leg", and sliding it along as the bag is filled.

Above: The bag is folded back to shorten the overall length, then unfolded as it is needed.

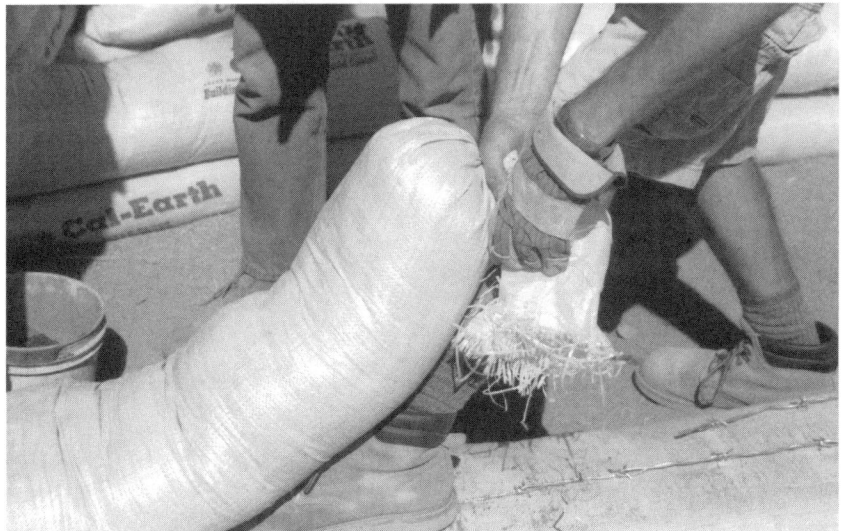

Left: Twisting the bag like a pony tail and tucking under to close the end.

SUPERADOBE - SANDBAG SHELTER

Right: Sometimes the end of the bag needs extra filling and packing by hand to avoid drooping, and to make an accurate line.

Doorway: Dome Entry

When a large opening is made in a dome, such as a full sized door, buttress walls are needed on either side. These walls are perpendicular to the dome, and will also make a protected entryway with a door vault over the opening to protect from rain. The buttress walls should be built at the same time as the dome so that the barbed wires and bags can connect together. Their weight resists the dome's outward thrust at the base, which is more intense on either side of the door opening. They are usually two to three feet long, or more, and are tied into the dome with barbed wire at each layer.

Below: The buttress walls may overlap into the dome every few rows.

Right: Continuing to fill bags for the dome and connecting them with the door buttress. Note the two foundation layers are two completed circles.

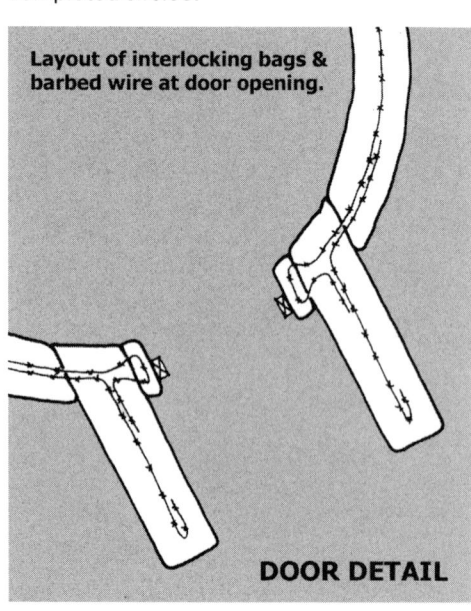

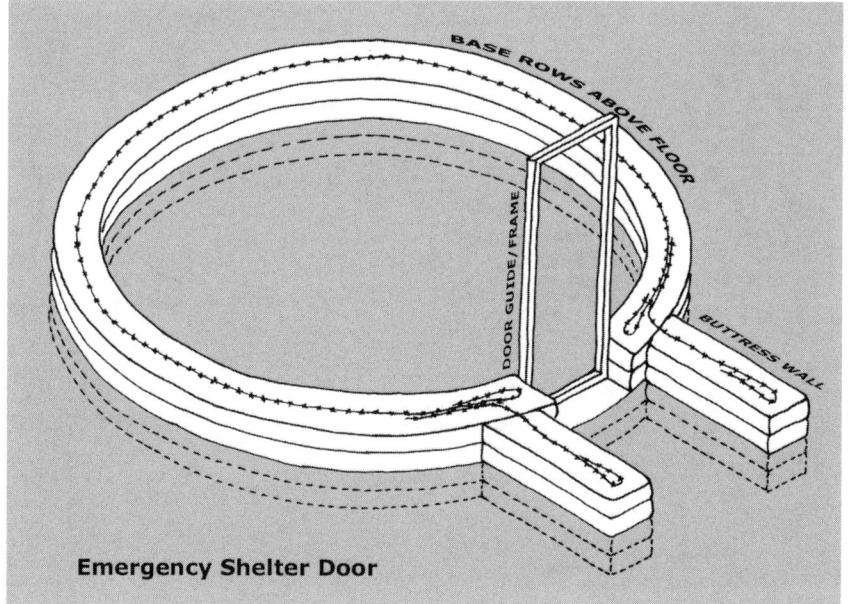

Emergency Shelter Door

Left: Axonometric view of the door frame set into the dome. Note the barbed wire connections for the buttress walls.

Setting In a Door Guide/Frame

It is useful to have a frame of available materials as a guide for the door opening, which should be set into the Superadobe base layers. If no materials are available, the opening can be built without a frame. The one shown is 2 ft. wide or 65 cm, and 6 ft. tall or 185 cm and has been tied into the bag walls using barbed wire. Since many emergency shelters are built after buildings collapse, doors or flexible materials such as carpets from among the rubble could later be attached to the outside of the door opening. Since the buttress walls are wider than the opening, the door covering need not fit exactly.

Above: Levelling the door frame can be done with a carpenters' level or a plumb bob (a weight tied onto the end of a piece of string).

Left: Using barbed wire to tie the frame into the dome wall.

SUPERADOBE - SANDBAG SHELTER

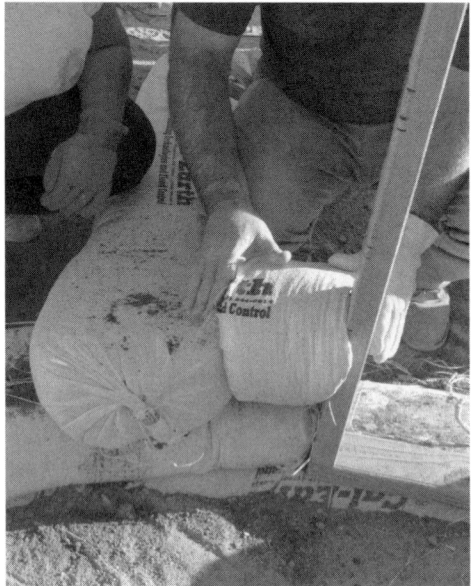

Above: A small bag is placed between the buttress wall and the frame.

Right (above): A metal plate helps the builder to slide the bag the up against the door frame without getting stuck on the barbed wire. After the bag is in position, the plate is removed. Note the height compass is low, at two bags above the foundation.

As more bags are placed and tamped up against the frame it will be firmly held in place. However, it should be temporarily braced open if the material is flimsy. If no guide or frame is used, wire loops or hardware can be sandwiched between the bags as door attachments or basic hinges.

Dome: Geometry and Compass

When the base is completed, you are now ready to start the dome which will gradually curve inwards to close at the top. The level where the base ends and the dome begins is called the spring line. At this level the dome needs a second compass to create the correct curve. This is called the "height compass" and it is positioned in the doorway on the outside of the foundation. Reading Appendix IV "Compass" is strongly recommended before continuing.

Below: The height compass can be raised up to a higher spring line. It is set into the ground or attached to the bags.

Right: Demonstrating how to use the center and height compass.

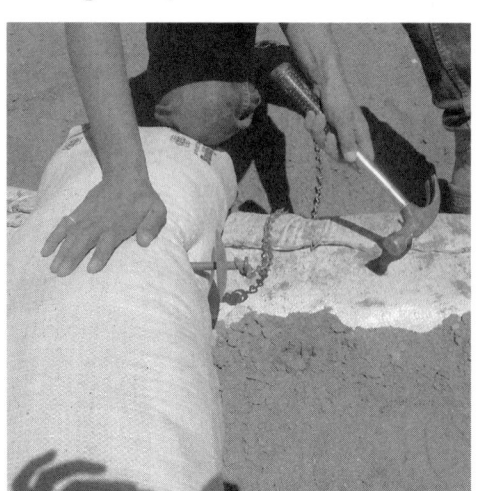

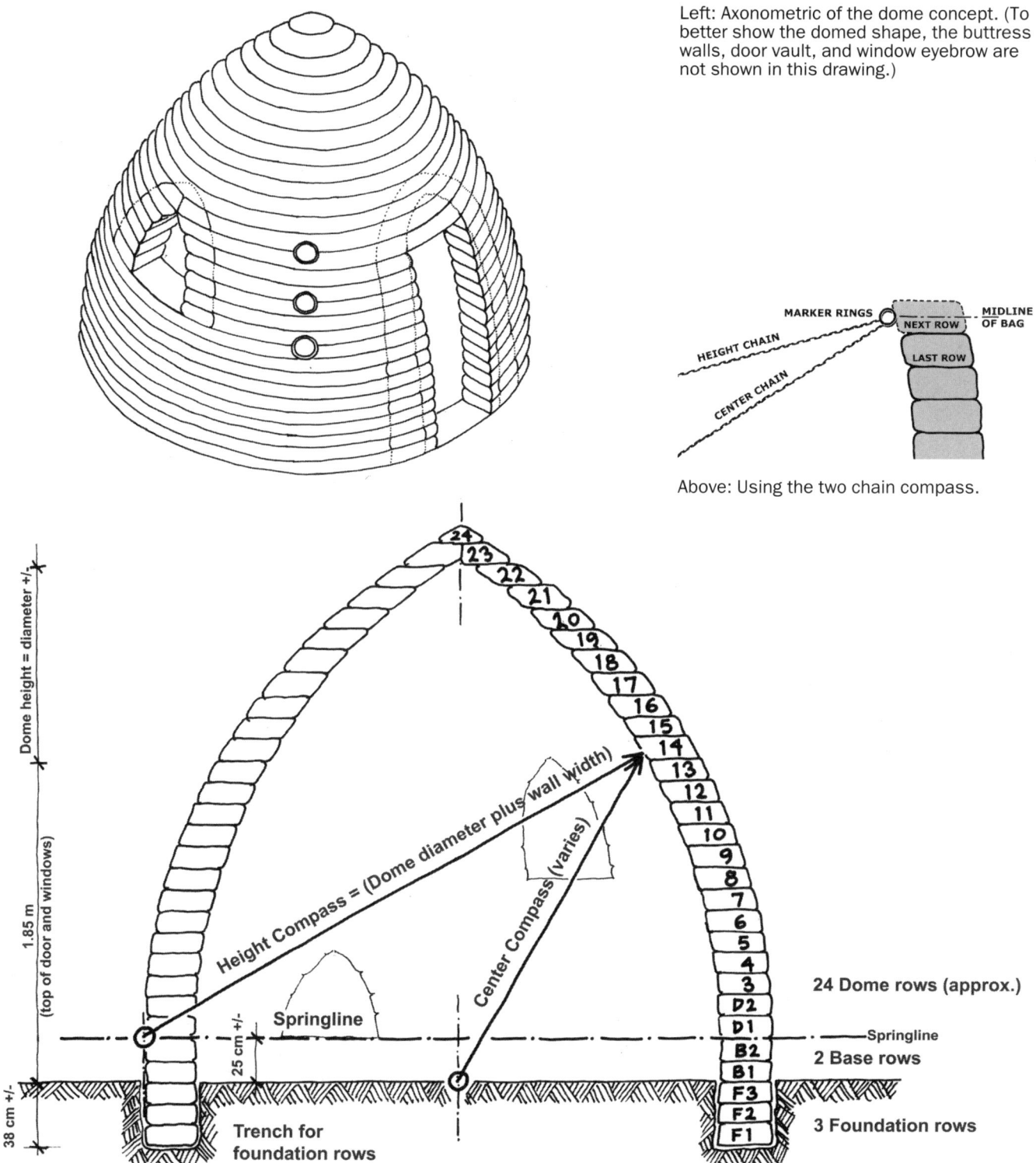

Left: Axonometric of the dome concept. (To better show the domed shape, the buttress walls, door vault, and window eyebrow are not shown in this drawing.)

Above: Using the two chain compass.

Emergency Shelter: 10 ft. / 3.0 m. interior diameter dome

(same for 8 - 12 ft. / 2.5 - 3.7 m range)

Compass section, Springline, and essential dimensions

SUPERADOBE - SANDBAG SHELTER

A LANCET DOME COMPASS can be made to build a Superadobe Dome, with two pieces of chain.

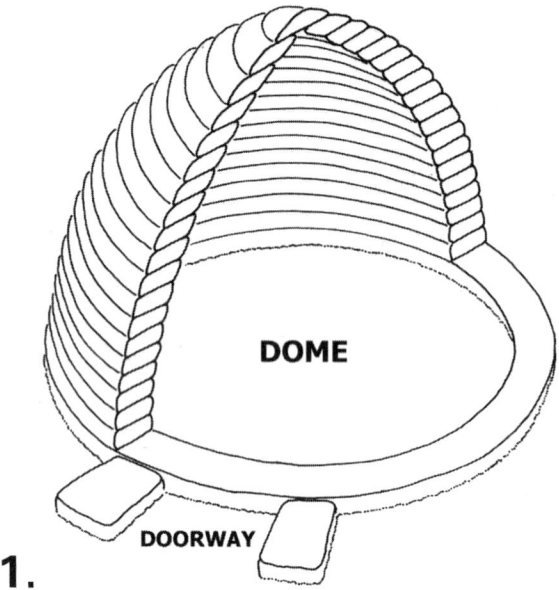

1.

A section view of a Lancet/Ogival Dome made from rows of Superadobe coils.

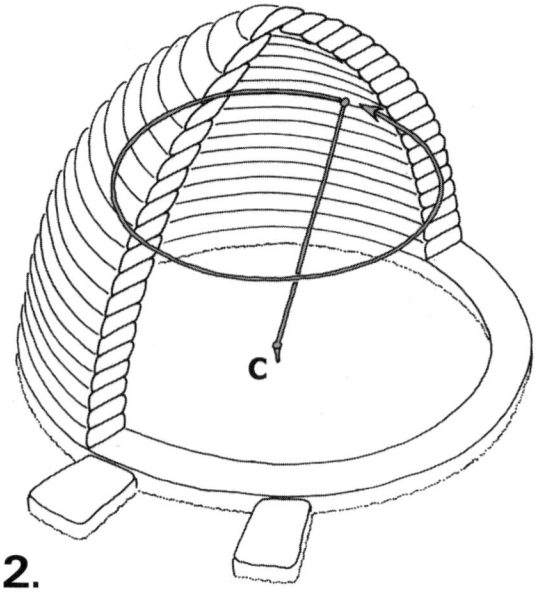

2.

The Center Compass (C) is the chain that rotates around the center and makes each Superadobe row. It gets gradually longer to construct higher rows. To know how long to make the center compass we use the height compass.

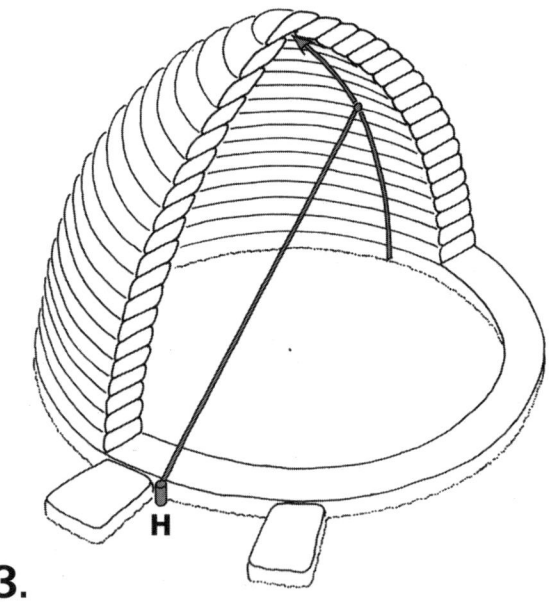

3.

The Height Compass (H) draws half an arch. It is the chain for controlling the shape of the Lancet Arch. It always stays the same length, which is equal to the dome diameter at the wall center. For practical reasons we fix the compass on the outside and draw the arch to the inside face. The height of the finished dome is approximately the equal to its diameter.

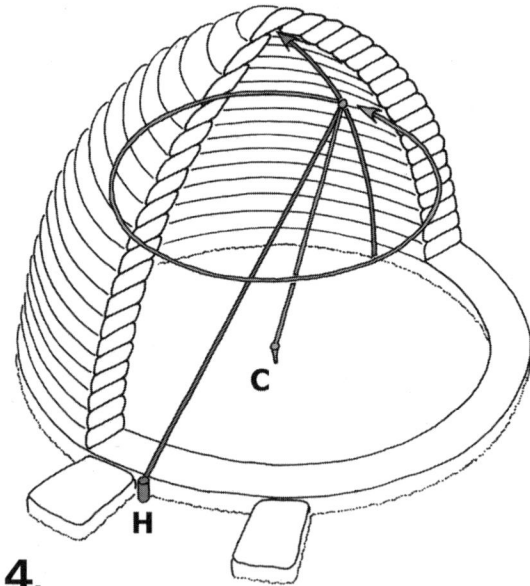

4.

As we begin constructing any row of Superadobe we bring the two chains together and adjust the Center Compass length by matching it with the Height Compass. Therefore, at every row, the Center Compass increases in length to match the Height Compass. Once the Center compass length is set, the Height compass is not needed any more for that row.

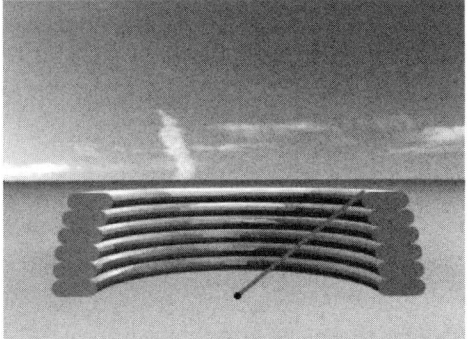

Above: Rows 4-6 the coils begin to step in according to the compass.

Left (above): The builder uses the center compass chain with his finger inside the marker ring. The center compass is lengthened at every row by pre-measured amounts, or by using a second chain.

Using the instructions from the compass chapter at the end of this book, from now on you will build ring after ring of Superadobe bags and barbed wire, tamping and measuring with the compass. Your compass may be a minimal two chains or ropes, or a steel pipe, or an entire frame assembly to guide the builder.

As the dome grows it becomes time to insert the windows.

Left: Children experience the size of the dome as it is being built.

Below: A steel pipe can also be used for a compass, with a bracket marker.

SUPERADOBE - SANDBAG SHELTER

Above: The essential elements for a pipe window are a section of pipe and a disc of glass or plastic.

Right: Assembled pipe window elements with openable glass.

Windows: Pipes and Vents.

For an emergency shelter, we must first ask the question, "How much light and air do we need?" And then ask, "Why can't we separate the two functions of light and air?" Larger openings for light do not need to open, eliminating the need for expensive window frames. Smaller openings for ventilation can more easily be made to open and close.

Therefore, for the view and daylight, a piece of glass or plastic can be directly grouted into a larger opening constructed with the bags, or fixed inside a larger section of pipe. For ventilation, short sections of pipe can be built into the Superadobe wall.

These small pipes can be opened and closed in an emergency using a wide variety of available materials; by stuffing them with cloth for example, or using items like storage jars, or clear plastic bags. Some of the pipes can be designated for light, and others used to control the ventilation.

Many pipe windows can be used in a single dome, depending on the amount and direction of the ventilation needed. These should be spaced apart by several Superadobe rows.

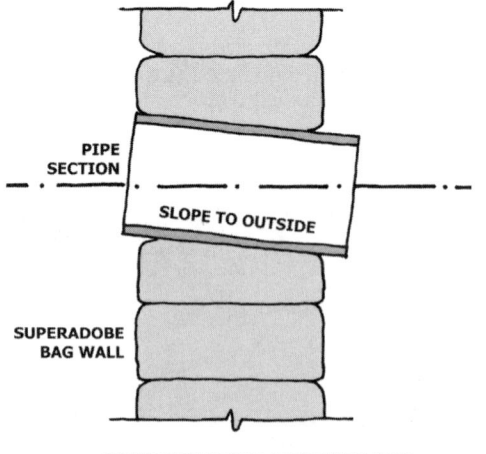

PIPE WINDOW SECTION "A"

PIPE WINDOW SECTION "B"

56 KHALILI - CAL-EARTH

Pipe windows and vents should always slope downwards and outwards to shed rain to the outside. The pipes should project outwards to protect the dome from rain, and also project inwards when needed to accommodate the curve of the dome.

Each length of pipe should be 2 inches longer than the thickness of the wall, and pipe can be anywhere from 6 to 12 inches in diameter.

Above: The bag is laid over the pipe window and shaped using a brick.

Left: The pipe is positioned to slope outwards.

Below: The position of the bag and the pipe is checked with the compass.

SUPERADOBE - SANDBAG SHELTER

Above and right: As the coils of bags begin to step in more, they must be continuously tamped and shaped with a brick or tamper.

Below: The center compass lengthens for each row (as determined by the height compass).

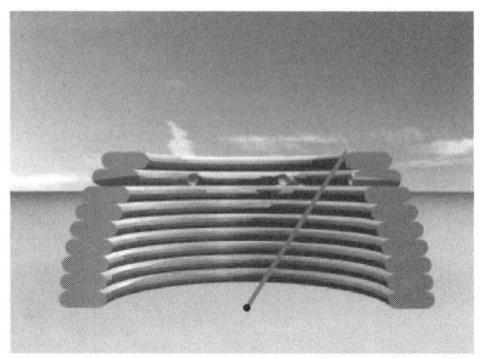

At first, the bags will undulate over the pipe windows/vents, but after a couple of rows they must be made level again, to allow the builder to follow the compass at all times. It is better to leave the bag empty going over the pipe, but fill the bag full on both sides; or omit the bag altogether over the pipe. The barbed wire must continue under and over the pipe.

Pipes can give the most beautiful lacy and patterned effect of light as well as giving great ventilation. For tropical climates you may need many pipes for a constant breeze; in desert climates a small skylight and pipes at the top of the dome will siphon off hot air; for cold climates a few lower level window pipes will provide extra light.

Above: After a few rows over the pipes the bags are level again.

In general, the pipe windows offer greater security than a large window (no one can climb through them, and in a war zone pipes laid at an angle may deflect bullets).

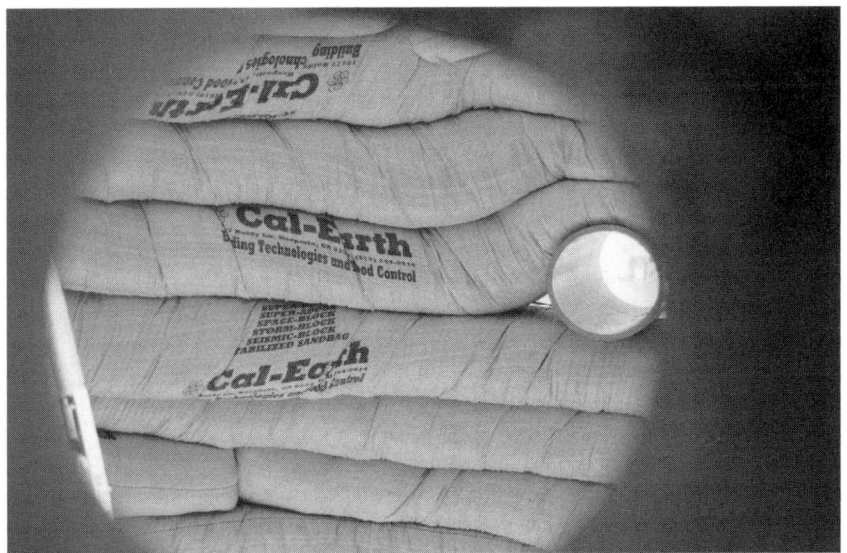

Left: Small round windows provide view, ventilation, and security.

SUPERADOBE - SANDBAG SHELTER

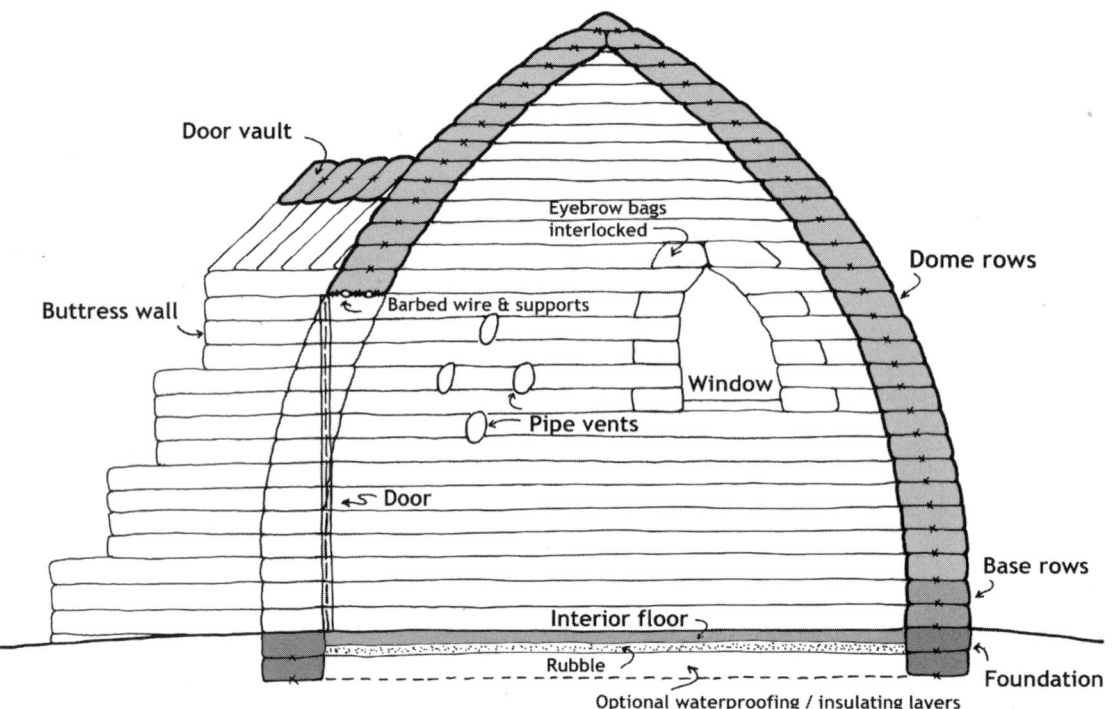

Section of Emergency Shelter: 10 ft. / 3.0 m interior diameter
(same for 8 - 12 ft. / 2.5 - 3.7 m. range)

Below: A short door arch is marked on the bags, which are cut with a saw and later removed.

Openings for Arched Windows and Doors

Many small round windows such as the pipe windows can be added with little or no experience by the builder.

For larger windows, or a small door, an unreinforced opening must be an arched, circular or triangular shape, because of the nature of earth construction.

The safest way to add arched openings is to leave plenty of wall area between them. When an arch is created in a wall, that wall itself is the buttress to support the arch. Therefore when the arch carries the load of the building above, the walls to either side are necessary. If there are too many large windows and not enough wall area in a dome, the arches will fail.

Arched Bag Form

To build a door or window opening, several methods are possible. For a quick opening, when no extra materials are available a "knock out panel" can be used. Imagine cutting the dome like a pumpkin lantern to make the window and doors. The knock out panel stays in place while the dome is being built, but gets knocked out when the structure is completed.

When the arches are tall or pointed, softer and less cohesive earths can be used. For shallower arches, the earth/stabilized earth material needs to be stronger to resist extra compression.

One way to make a knockout panel is with a hand saw, cutting the bags and leaving the pieces in place. As you build up the dome coil after coil, the bags can be marked with the line of the future window or door. If the earth is stabilized with cement or lime you will need to make the cuts while the stabilized earth mix is still wet (and insert a piece of bag material, like dental floss, to stop the cut from closing). If the earth is not stabilized, the arch may be cut at any time. The barbed wire is omitted where there will be an opening.

When using these methods for a larger door, remember to also build the buttress walls on either side of the door opening before removing the knockout panel. Windows or small doors situated in a thick wall do not need extra buttress walls.

Sometimes a knock out panel is part of a long term design and is left in place for a future opening, such as a door into another room. This method is especially useful when automated methods such as pumps are filling the Superadobe coils. The planned opening must work structurally both while closed and when it is opened up.

Above: Stabilized earth bags are cut while the earth material is still soft enough, using a saw or steel wire.

Left: A pre-cut knock out panel is removed.

Below: Several cuts are made at angles to more easily remove the blocks, and the bag material is used as a "floss" to keep the cuts open.

SUPERADOBE - SANDBAG SHELTER

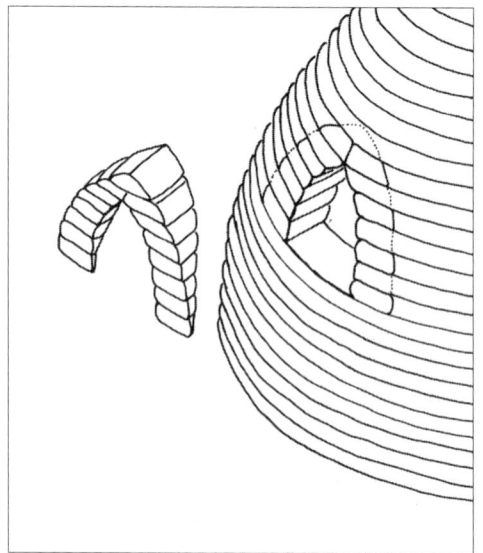

Above: Window arch and "eyebrow" concept.

Right: Exterior view of the arched window and bag form, where the bags themselves are used as the temporary arch form. The eyebrow is made with short bags which are set perpendicular to the dome wall.

Below: A form is made of bags without the barbed wire, which are filled with dry, untamped earth. The bags are stacked vertically.

Another way to make a knock out panel uses earth filled bags without needing a saw. The arched panel is built with separate earth filled bags which are loosely filled and not tamped so that they can be pulled out and emptied after the dome is completed. This method is practical for smaller arches.

Eyebrows over openings

Since the dome wall is curved, any window which is open to the exterior needs an overhang or "eyebrow" over it to protect the interior from rain. When the arched opening is formed over a vertical stack of earth filled bags, the eyebrow can be made by corbelling small bags out from the dome wall at every row. Other ways to create "eyebrows" can be seen in the dome variations later in this book, including long bag eyebrows.

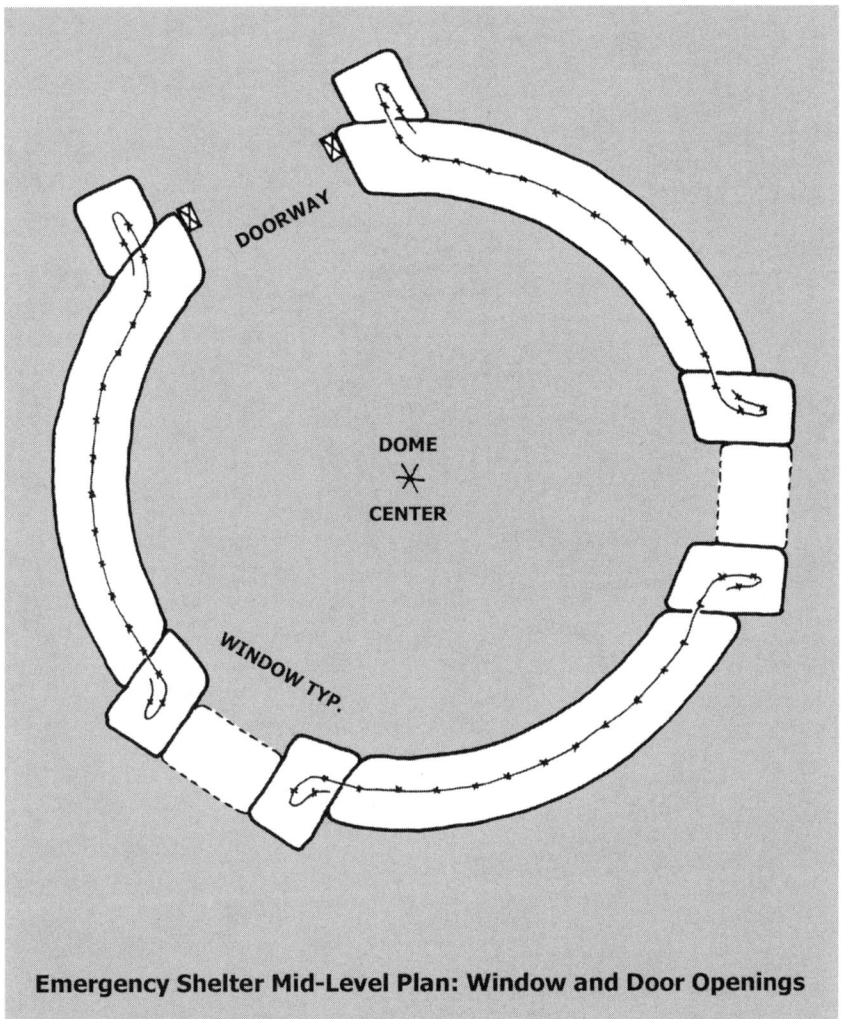

Emergency Shelter Mid-Level Plan: Window and Door Openings

Above: Building the window arch with small bags over the bag form.

Below: View from the dome interior. The walls are built up around the window arch before the bag form is removed.

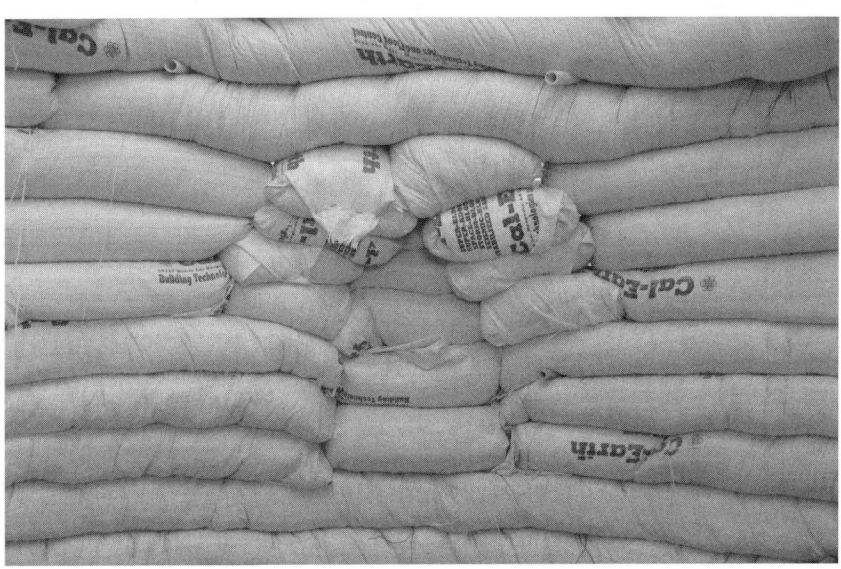

Below: Tamping the last two bags over the form to complete the window arch.

SUPERADOBE - SANDBAG SHELTER

Right: The bags have been pushed out to leave the window opening. This is possible because the barbed wire is not carried across the window opening, and the bags making the temporary form are filled with dry earth. These can now be reused.

Below: Interior view of the window opening after the bags have been removed.

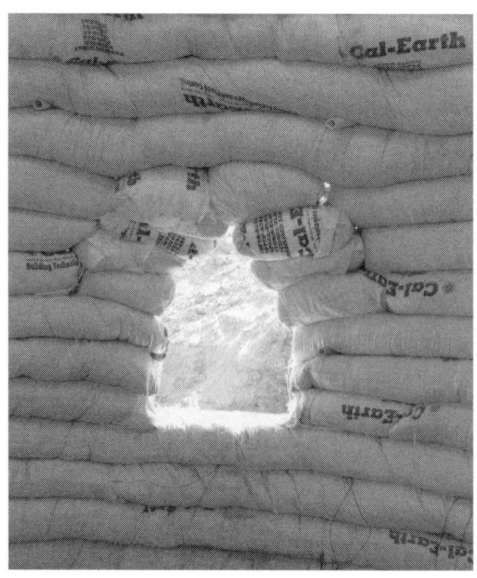

When you have built at least four finished, dried and set rows of bags above your door or window arch, you can cut the bags open, or push out your temporary form.

In emergencies, if there are no windows available and the weather is harsh, the inhabitant can leave the bag form in place. Later when the weather is warmer, the bags can be removed to open the window arch, and later a frame with glass can be inserted.

Reusable Form

If materials are available, a removable form can be made of plywood, metal or other material which can be reused for many shelters. It supports the Superadobe bags and makes it easier to create the arch and eyebrows over the windows and doors. In some cases the form remains in place as the window or door frame.

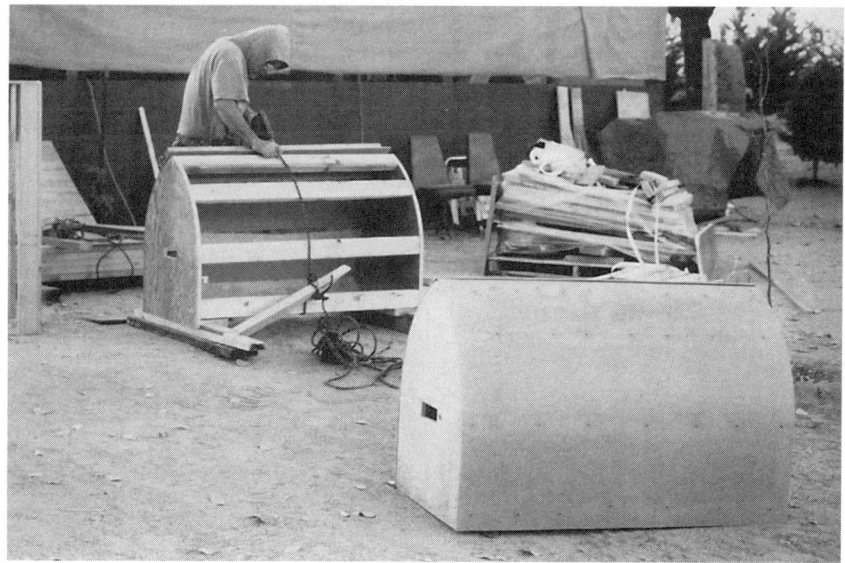

Left: Making wooden arch forms, with a substructure and plywood or sheet metal finish.

Above: An oil barrel or other as-found object can become a temporary arch form. Damp earth is packed around and over it to make the desired arch shape.

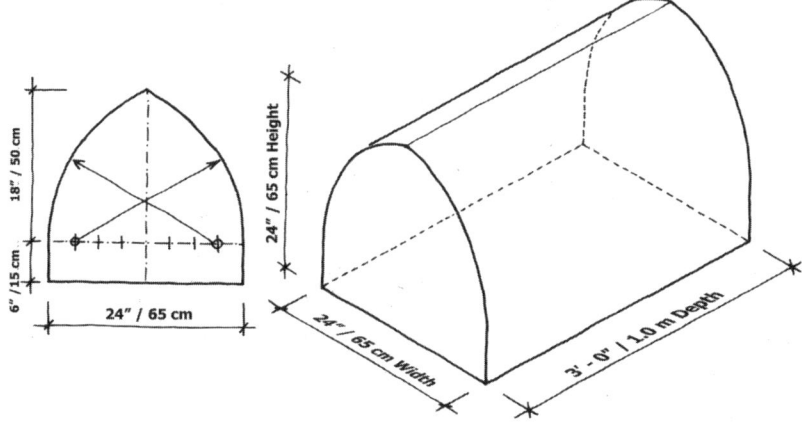

Form for Doors and Windows (typical small shelter)

Below: The oil barrel packed earth are being removed to leave a taller arched opening.

The arched form should be deep enough to support the dome curving inwards, and the "eyebrow" stepping outwards. A suitable form for a dome of between 8 and 12 ft. diameter (2.5m to 4m) is 2 ft. wide (65 cm) and 3-4 ft. deep (1m) for windows, door and niche openings. For a 10 ft. diameter dome or less the openings should not be over 2 ft. wide (65 cm). Other examples of forms are illustrated in the shelter variations.

A good form is lightweight yet supports the bags without deforming during construction; it must be strong enough to resist the impact of tamping. It is raised up on wedges, and should have a handle to be removed and reused without damage. Most forms need to be oiled and protected from moisture for a long life.

In fact, any available materials from an oil barrel packed over with damp earth, to a frame of steel bars, or the bag roll itself can become a reusable form.

SUPERADOBE - SANDBAG SHELTER

Stages of constructing a window (or door) with corbelled eyebrow.

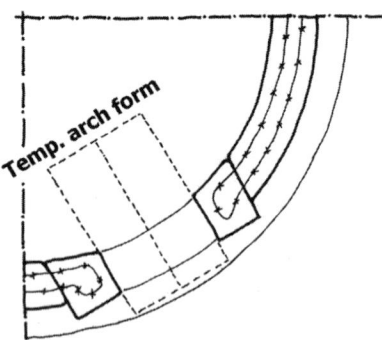

1. First row for windows (or door typ.) Small bags are placed along the form.

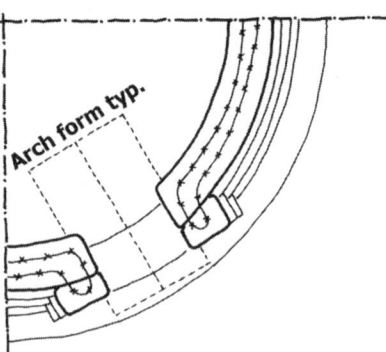

2. Middle rows to build "eyebrow". Alternate this layout with #1 layout to interlock the small bags.

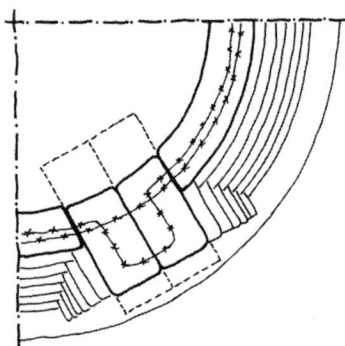

3. Top row closes over the arch form. Place a barbed wire row across all bags.

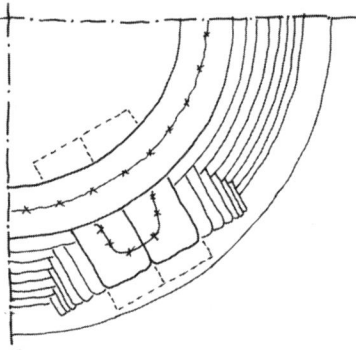

4. Continue rows above window/door as complete coils.

Left: The arched window opening after the form is removed.

Below: Arched window before the form is lowered and removed.

Extra barbed wire around openings adds necessary friction and cohesion. The barbed wire, for every window or door, is either doubled back near to the opening, or is continued up into the next row so that the opening is reinforced.

The bags must be neatly tucked under or pinned at their exposed ends, and tamped with a brick to become compact and strong.

A reusable form helps the builder to repeat a precise shape with the bags, which is desirable when fitting manufactured windows or doors into the opening. The form is removed only after at least four Superadobe rows are completed above the arch, or when the dome is completed. The form can also be made so that a part of it is left in place to become the permanent window or door frame.

As a rule, at least three to four feet of solid wall must exist between major openings. Therefore a small dome of 10 ft. diameter (3m) with openings of up to two feet wide could have a maximum of two windows and one door. The small round pipe openings are exempt from this rule and can be more plentiful.

Above: Starting the bag a little further back makes a stepped wall.

Right (above): An entryway built over a form can be made like the windows. It continues the logic of the stepped buttress walls.

Below: Using the steps to climb and build.

Right: Exploded axonometric of the relationship between the door opening and door buttress.

If a form is used for the entryway it can be built with small bags like the windows. The eyebrow element can be made deeper for a more protected entry. Without a form, the door is made as follows.

Buttress Walls:

The door buttress walls stabilize the doorway structure and also make a protected entryway. The door buttress bags connect to the dome with barbed wire and by overlapping into the dome wall. During construction the buttress walls can be stepped back and used as a staircase to climb up and down.

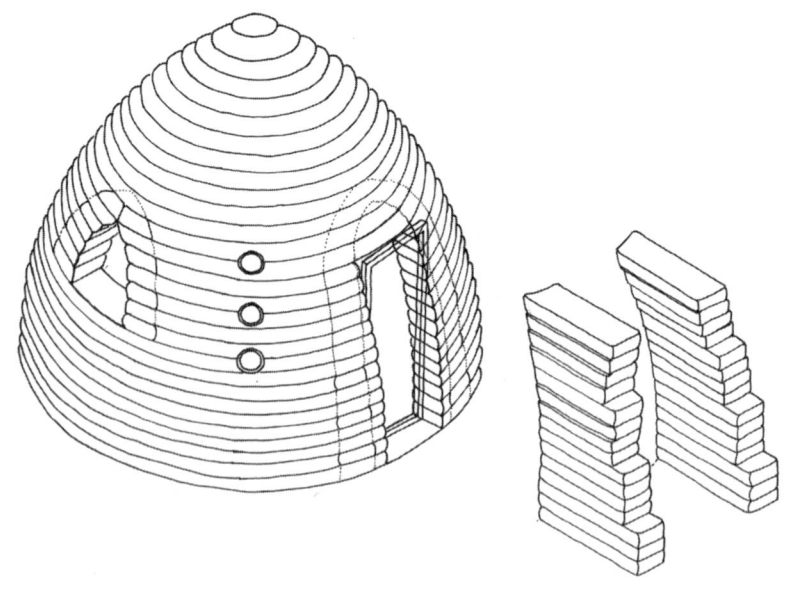

Left: Inside the door vault a small area of the dome wall shows next to the door frame. If a standard door and frame is not available, this area of dome wall can serve to attach an alternative fabric door covering, or "soft door" (see Model 8: Sinapsoapsis).

When no form is available, the door opening must be 2 ft. wide (65cm) and the buttress walls 3 ft. apart.

If the door frame needs to be mechanically attached to the walls, barbed wire can be nailed to the frame and sandwiched between the bags. The frame must fit snugly into the dome walls, as was shown in earlier photos. If needed, extra bags can be packed around the frame to close any gaps.

For a flat lintel over the doorway, extra barbed wires should be tied into the outside of the dome/buttress walls from several rows lower than the lintel, to support it. These wires must be at least six feet long, and will support the lintel from below.

Below: Filling gaps between the frame and the dome wall.

Left: General view of dome and door buttress.

SUPERADOBE - SANDBAG SHELTER

Right and below: Several strands of barbed wire are laid across the door opening. These must be connected into the dome several rows below the present layer.

Going Over the Door

In order to fit a standard rectangular door, we can build a Superadobe unreinforced lintel over a small opening of up to two feet span (65 cm). The bags can span this short distance if they are:
 a) filled with stabilized earth,
 b) supported from below by several barbed wire strands, or
 c) supported from below by as-found materials (e.g. a short plank of wood, pieces of steel pipe or steel reinforcing bars).

The challenge is to make the first row over the doorway, which needs the most support especially during the tamping. Here, the trainee

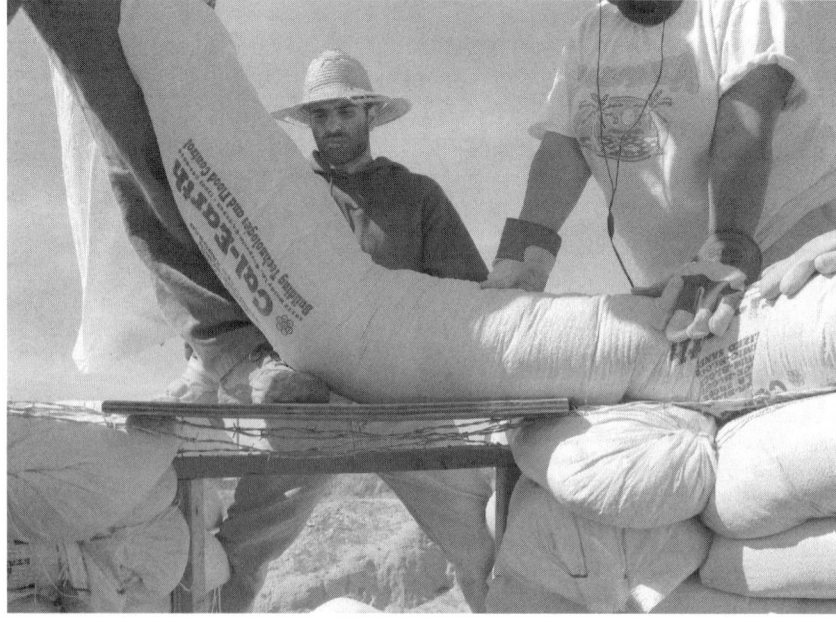

Right: The filled bag is being laid over a short plank; barbed wires are fixed above and below it. The plank can be later removed, or left in place.

Left: The first two bags have been completed and tamped over the doorway.

builders filled and placed the long bag over a short plank of plywood with three strands of barbed wire above and below. After it was placed, the bag was tamped as usual to compact the stabilized earth and gain strength.

If we look at the lintel bag as an integral part of the whole dome, we can see that the rows above it actually help to carry the lintel row because of the barbed wire tension element. In effect, each succeeding row makes a thicker and thicker beam element. These rows above the door have two strands of barbed wire, to provide extra resistance to tension, which should be well sandwiched between the rows of bags before tamping.

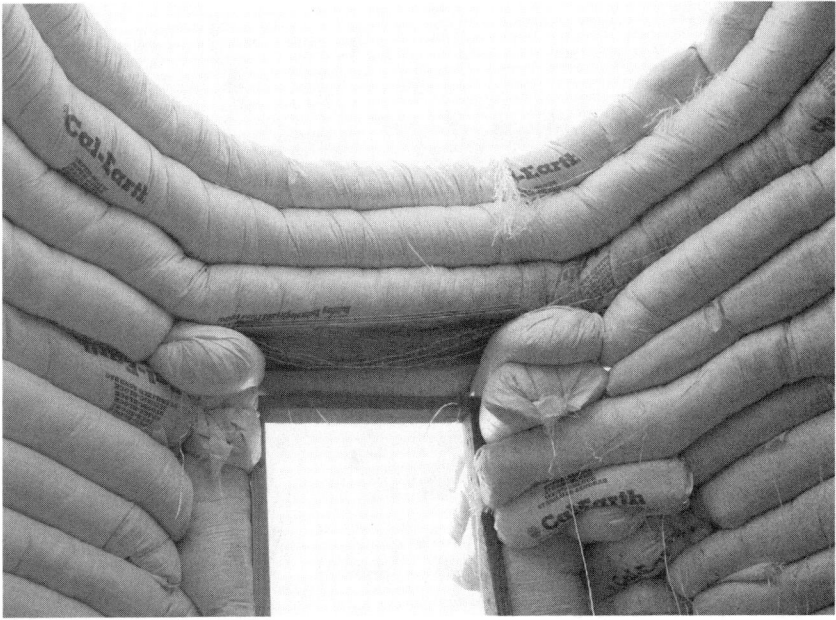

Left: The dome continues as normal above the lintel.

SUPERADOBE - SANDBAG SHELTER

Above: The top rows step in faster.

Right: Standing on the coil steps, below the top of the dome.

The Upper Part of the Dome

As you build up towards the top of the dome you will need to refine your technique to allow the rows of bags to step inwards more and more. "Corbelling" is the technical term for this stepping inwards. You may be surprised how much the coils will need to corbel inwards.

By applying the following steps, even first time builders in training have built strong domes.

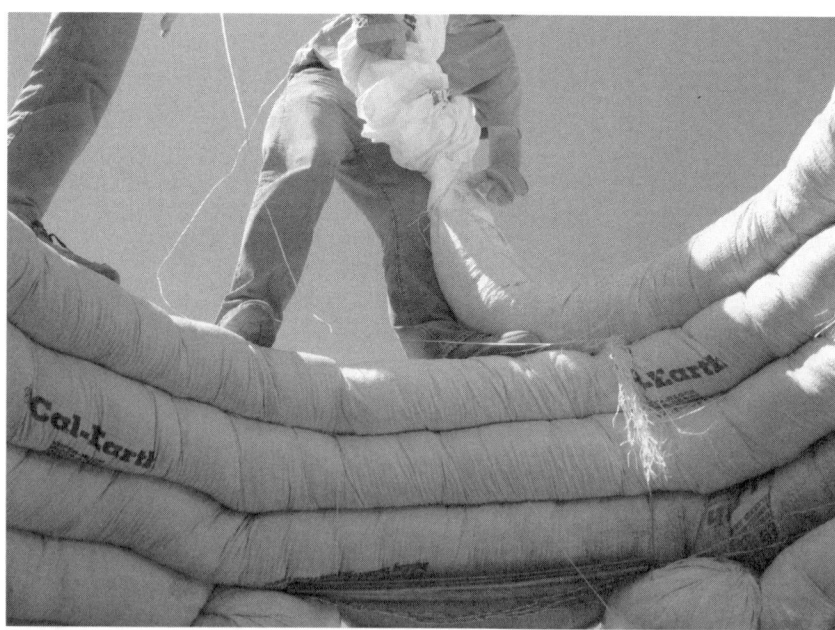

Right: Twisting the fabric of the bag as it is filled, to create a smooth curve.

Above: Tamping with a slight slope towards the outside.

Left: Filling and coiling the upper rows.

1) When you walk on the dome, step on the outside edge of the bags. Try to stand a couple of rows down from the top. Previous rows may still be soft if you are building quickly.

2) At first, set the bag directly over the lower row. As you fill and twist the coil, the bag will naturally work inwards towards the compass line.

3) Start shaping the bag with a brick right away, three or four feet after it is filled, to stop it from slipping too far while you build. Keep measuring with the compass, and if needed fill the bags with less earth to make them flatter and wider. It is easy to tamp the bag inwards, but harder to push it outwards.

4) Be sure that the finished bags follow the compass curve. Don't be afraid to remove bad work since it is far better to lose one row of work than to leave an unsafe base for the next rows above.

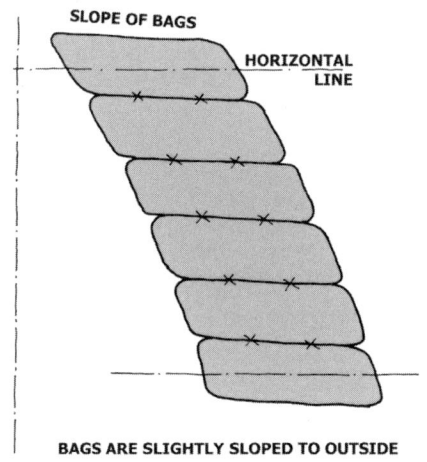

SUPERADOBE - SANDBAG SHELTER

Above: The upper layers of bags are first laid in a slightly wider circle, then tamped inwards to meet the compass.

4) Tamp the outer part of the coil first. Remember that you only have supporting bags underneath the outer part of your coil.

5) Tamp to give the bag a gentle slope towards the outside; it will help the next row on top to step in more, and will help to shed rainwater off the dome.

6) Use a brick to shape the bag from below if you want a smoother curve on the inside of the dome.

7) Place the barbed wire towards the inside edge of the bag and make sure that it is gripping both above and below.

Windows should not be built on the upper rows by inexperienced builders. Small inserts may be sandwiched between the bags (such as vents, sticks, rods, or wires) for scaffolding attachments or roofing which will be added after construction.

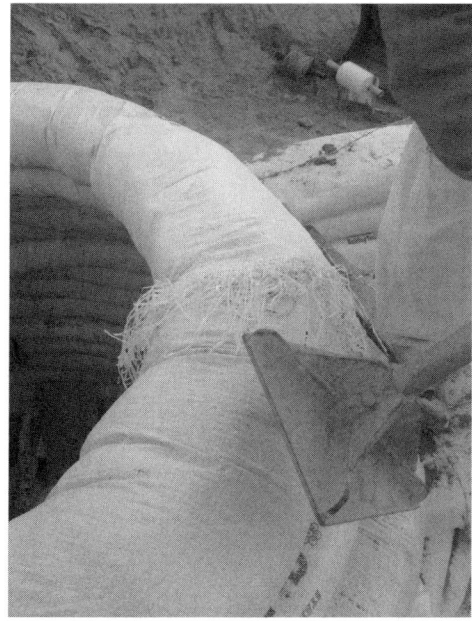

Above and left: The coil is tamped while being checked with the compass. The tamper pushes the bag inwards to meet the compass line.

Left: The builder uses the coils as steps and does not stand on the soft upper bag until after it is tamped.

SUPERADOBE - SANDBAG SHELTER

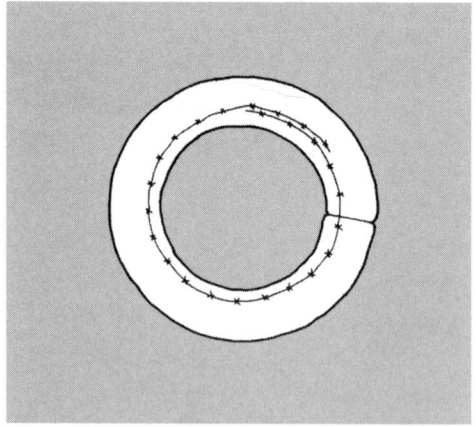

Above: Plan of a typical upper layer above the door and windows. These layers have continuous bags and barbed wire to tie the dome together at the top. This is important to the overall stability of the dome.
Note that the barbed wire is joined by overlapping or twisting together where the bag is continuous, and the bag is joined where the barbed wire is continuous.

Right: Coiling the bags at the top of the dome.

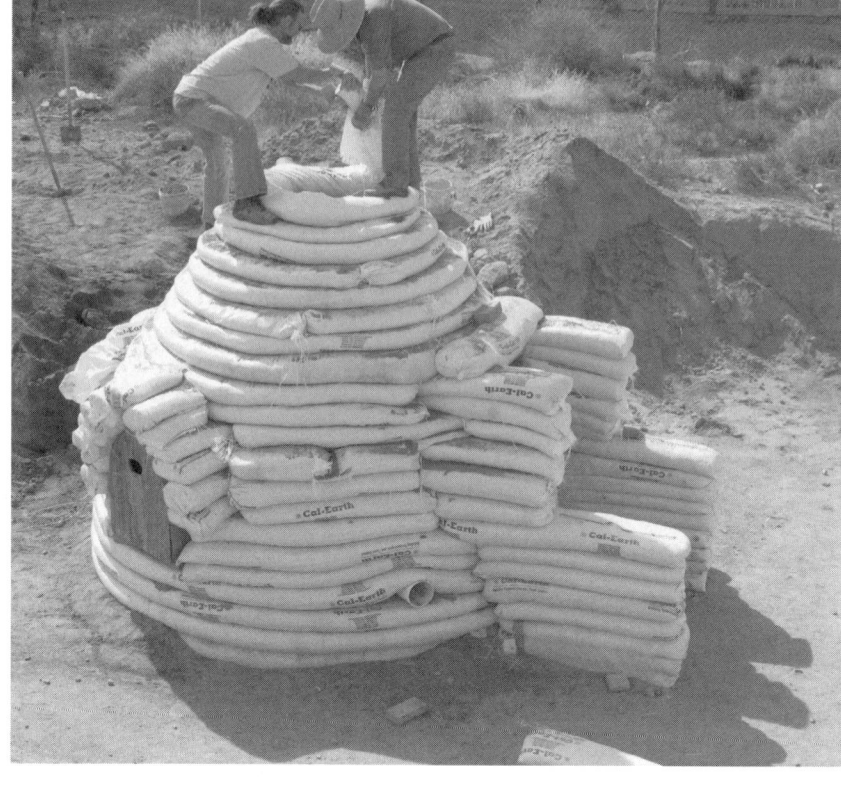

Right: Curving the bag and shaping/tamping with a brick at the same time.

Below: Section through the dome showing the corbelled upper layers.

8) When the curve gets really tight the bag should be filled and shaped simultaneously, using a brick. This shaping and sculpting helps to stop the bag moving around, and keeps it firmly on the curve until the whole row is tamped. Filling and shaping can be done by one builder, or by two persons. An experienced builder can build the last rows with a spiral bag rather than individual circles.

76   KHALILI - CAL-EARTH

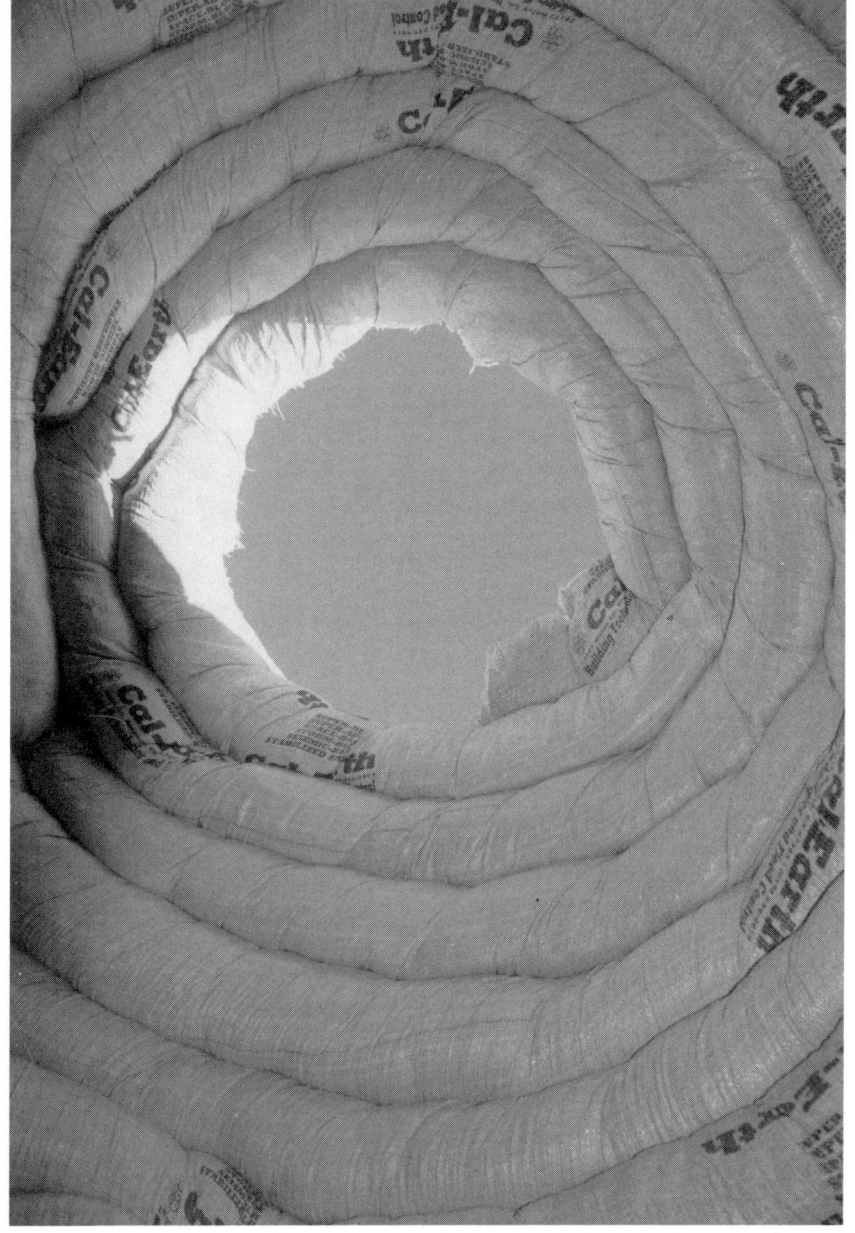

Above: All the coils of the dome are completed.

Left: Looking up inside the dome as the coils are almost complete. Here the top rows are being built as a spiral.

Below: Very small rings close the top.

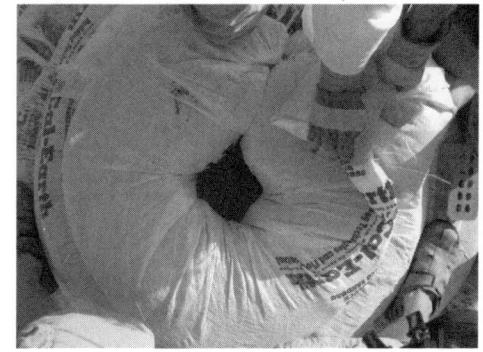

The beauty of the organic domed form is how completely it works in harmony with nature and gravity. The lower part of a dome consists of wide, curved coils which can only step in a little bit, and this is all that is needed at this stage of the dome. At the top of the dome, when the coils must step in as much as three to five inches, the curve is so tight that each coiled ring is partly self-supporting. These tightly coiled rings are easier to build with 14 - 16 inch wide bags. If the lower part of the dome uses wider bags you may decide to close the top by using a smaller bag.

As you build up to the top, you should decide if you want to close it completely, or leave a skylight for more daylight.

SUPERADOBE - SANDBAG SHELTER

Above: Model to show triangular doorway is cut leaving the base bag intact.

Right: Pushing out the first block of the sawed coil.

Right (below): The door opening can be cut up to 5 ft. high. This photo shows a minimum crawl-through opening, like a tent of similar size, 3 ft. 6 inches wide at the base and 1 ft. wide at the top.

Below: Cutting out the rest of the door.

Alternatives to the Door and Lintel: Cutting Open the Door

Gravity and the strength of the earth/stabilized earth material decide the size and number of the openings. If your dome is small and temporary you may decide to quickly cut open a door. The opening can be cut up to 5 ft. high. However, the door should never cut into the bottom two foundation rows.

Left: Lower or higher doors can be cut from the side of the dome, like a tent. The foundation rows are not cut.

For any opening, the weight of the dome above must be carried to the ground through the mass of the walls, by travelling around the opening to reach the ground. There must be at least four feet of wall mass between openings, and doors and niche openings should be seperated as far apart as possible .

Cut out door or window openings should be arched, triangular, or lozenge shaped like the photo. Only small cut-out openings can be made without buttress walls. To protect the opening from rain a tent-style covering may be added. Either sandwich the tent-style covering into the row above the opening and let it hang down to cover the door or window, or sandwich an eyebrow projection such as a piece of wood, plastic, or metal, and later attach the tent-style covering beneath it.

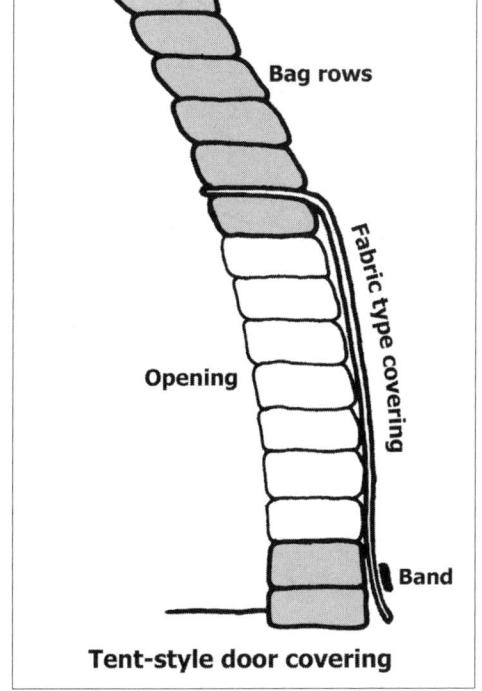

Tent-style door covering

SUPERADOBE - SANDBAG SHELTER

Right: The door and lintel are completed. (More lintels could be added to make a "mineshaft" entry, if the earth is stabilized.)

Entry Vault:

A small vault, added over the top of the buttress walls, protects the door opening from wind and rain. The traditional technique of leaning arches can be used to span over the entryway, eliminating the need for a temporary form.

When the door and lintel are completed, a leaning arch can be built of unstabilized or stabilized earth. Learning the leaning arches technique also enables the builder to create larger entryways and connecting vaults between several domes.

Above: An entryway could also be made starting with a window opening, then cutting down to the floor and removing the bags below the window opening to make a door. Be sure to leave the bottom two foundation rows.

Right: First, two small bags are packed leaning against the dome.

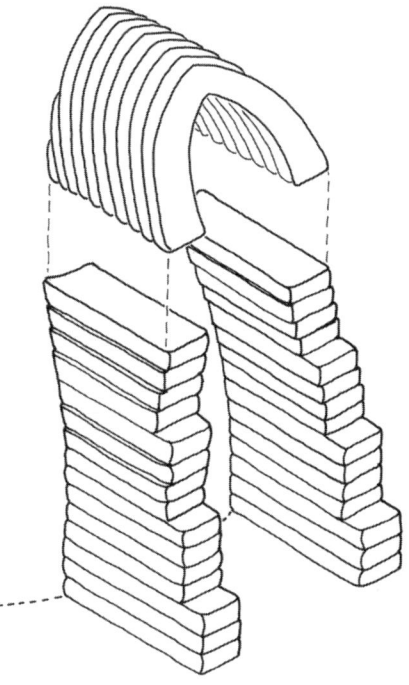

Left: The drawing shows the door vault of leaning arches, and the two buttress walls which will support it.

To build a leaning arch, first two short bags are set on the buttress walls and pitched against the dome. Then two longer bags are shaped over the small ones. And then longer ones until the two sides meet and form an arch.

If leaning arches cannot be learned in an emergency, and if stabilized earth is available, a series of short lintels could be quickly added to make a "mineshaft" style entrance, spanning across the buttress walls.

Left: Next, two longer bags are shaped until a leaning arch spans over the entry.

SUPERADOBE - SANDBAG SHELTER

Right: Elevation of the dome entry showing the relationship between the door and the leaning arch entryway.

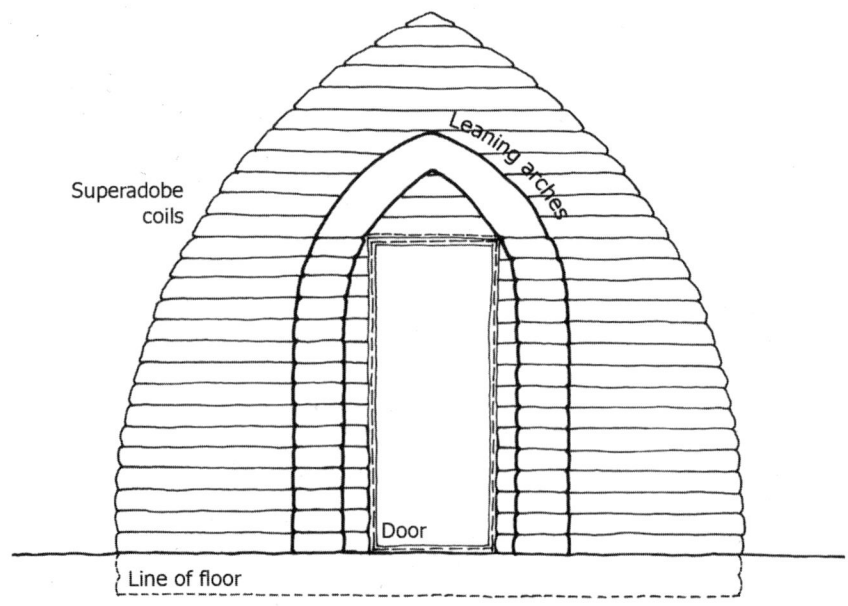

DOME ENTRY ELEVATION

Repeating the leaning arch makes the vault longer or shorter as needed. While the bags for the arch are being filled with earth, they must be continuously shaped and tamped with a brick, so that they maintain the arch shape for both sides to meet in the center. The barbed wires connect the upper layers of the buttress walls and the leaning arch bags, and are placed between each leaning row.

Right: The leaning arch used here is a pointed arch made from two filled bags, meeting at the center.

Left: Entry with leaning arches.

Each element of the entryway (the opening in the dome, the door and frame, the buttress walls and vault) is somewhat flexible in relation to the other elements. For example, depending on the dome size, the door and frame can be set into the dome wall or set into the buttress wall and vault. The opening in the dome can have a small flat lintel or be arched. However, the overall relationship of elements must be consistent with the dome size and material strength.

If a larger, wider entrance is desired, extra buttress walls must be built to support the larger and heavier arches above. The dome below is an example of a wider entry vault with added buttress walls creating a curved bench and landscaping. A wide entry could become a connecting room between two shelters.

Left: A larger arched entryway with buttress walls which was built over a form.

SUPERADOBE - SANDBAG SHELTER

Right and below: Fondation pour L'Architecture exhibit, Sandbag shelter built by Cal-Earth and NEST (New Start Trust) in collaboration.

A Well-Built Sandbag Shelter in Belgium

This single dome with a wide entry vault was built to demonstrate the emergency shelter concept at the "Fondation pour L'Architecture" for an exhibition in Brussels, Belgium.

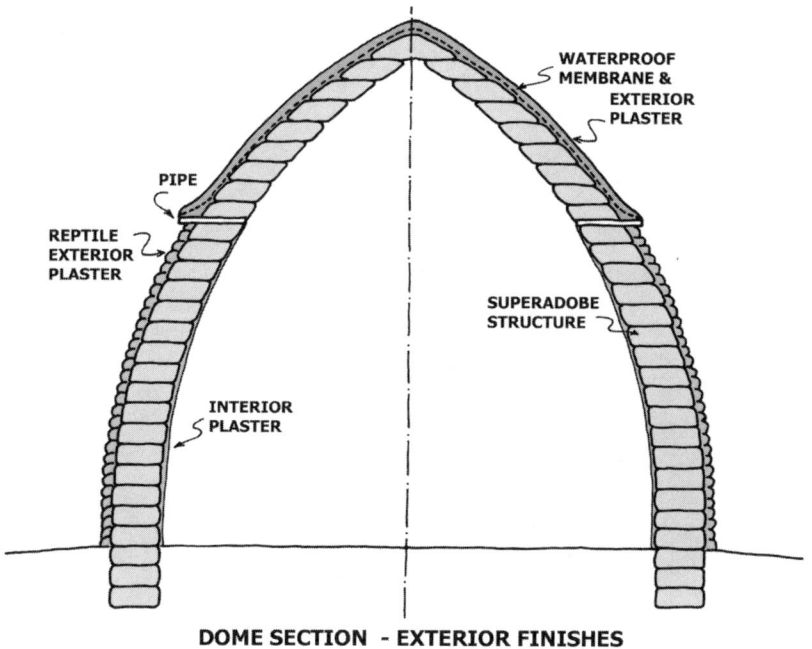

DOME SECTION - EXTERIOR FINISHES

It shows how, with some practice of the techniques, your first-time training dome can be improved to achieve this nearly perfect symmetry. The bags are laid in dense, even rows and neatly finished at the ends.

Waterproofing and Finishing

After constructing your sandbag shelter, it will need protecting with waterproofing and finishes, unless you are in a very dry desert.

For quick waterproofing in an emergency, a tarp or plastic sheeting can tied over the dome. Embedded attachments such as small pipes, short wires or rope are sandwiched between the coils during construction and the sheeting is tied onto the dome with these.

For long-term waterproofing and finishing, it is best to follow local practices and materials suitable to the local climate.

Examples of long-term finishes include:

1) Membrane type waterproofing. These are usually applied in combination with an exterior plaster, and are commonly a liquid or semi-liquid, viscous material, which is applied onto an exterior plaster, or mixed with it. For example, a scratch coat of plaster is first applied over the bag structure to smooth out the surface, then a waterproof membrane, such as layers of asphalt and fabric, is applied over this, and finally a protective surface such as stabilized earth plaster may cover the waterproof membrane.

2) Discrete elements. This includes a range from traditional dome finishes such as metal or ceramic tile, all the way to innovative adaptations of traditional roofing covering such as tiles, shingles, corrugated steel, thatch and palm leaf palapas, which may be anchored onto the dome using the embedded attachments above.

Several examples of finishes are described later in this book.

Above: A sheet of plastic can provide immediate temporary waterproofing.

Below: Rope or cable may be threaded through the small pipes sandwiched between the bags, to tie on a quick waterproof plastic sheeting or other long term roof covering.

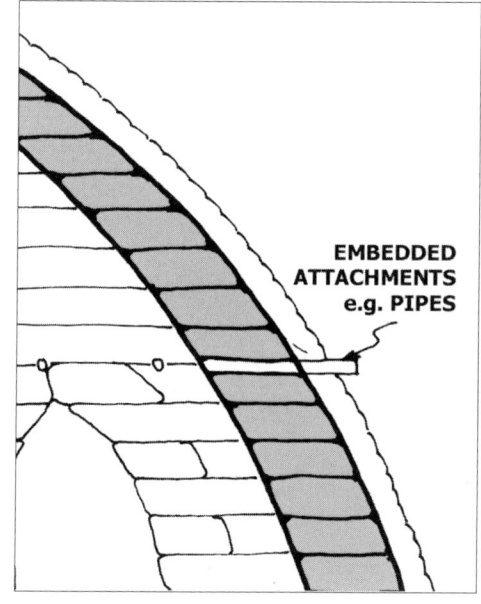

SUPERADOBE - SANDBAG SHELTER

Above: The class tests the size of a 10 ft. diameter shelter, in their 15 minutes break. A dug out pit, a bunch of small buckets, bags, and barbed wire, can do miracles in creating safe and beautiful shelters.

Reviewing What Has Been Learned

Now that this first training dome is finished, built by trainees who had never built one before, you can see that with no previous experience and despite a few imperfections, a successful shelter has been constructed.

Let us review what has been learned so far:
1) Filling long or short sandbags and coiling them into a dome.
2) Using a compass to make the correct domed shape.
3) Inserting pipe windows and vents.
4) Making door and window openings.
5) Using leaning arches to make an entrance.

Now it is time to get more hands-on knowledge by following the design variations and combinations of the basic Emergency Shelter. To truly understand how gravity works with structure and form, the best way is to keep building small 8 - 12 ft. (2.5 - 3.7 m.) inside diameter domes. The ones shown are built as prototypes at Cal-Earth Institute and begin to address the issues of variations in climate and human needs, including upgrading a temporary shelter into a permanent structure, while demonstrating a higher level of builders' skill.

*"earth architecture is like a choir
repeating the same music, same theme
its strength is in its unity
its beauty in its repetition"*

C·H·A·P·T·E·R 2

ECO-VILLAGE
AND CLUSTERING

SUPERADOBE - SANDBAG SHELTER

Above: Single dome unit concept.

Right: Repeating and clustering single domes.

Design and Technique Variations

A single domed emergency shelter may be repeated and modified so that it becomes the basis for unlimited variations in design and technique. In just the same way a human face, with small changes to its main features, becomes every different nationality and character. The best way to understand variations in design is to learn about them at the same time as some variations in construction techniques.

These variations create a new language in design and construction. Three essential concepts in this new language are:
1) Clustering
2) Endless Wall
3) Courtyard.

Right: Repeating and clustering domes under construction (unfinished Hesperia Museum project).

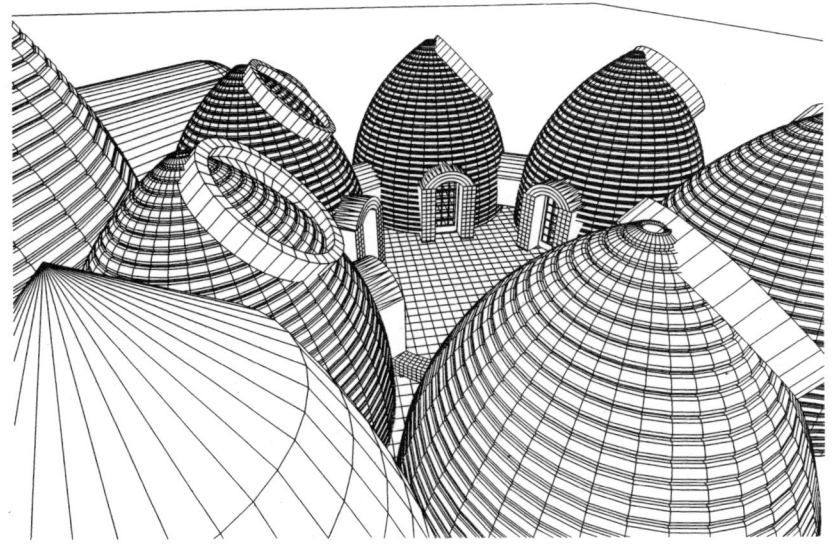

Above: Variations of organically curved seashell forms are the inspiration for structures and spaces.

Left: Domes clustered around a courtyard for the Hesperia Museum design.

Clustering

Using timeless principles in nature, such as the geometry of sunflowers, gravity, the proportion of the human body, wind, sun and shade, small and large shelters can be clustered into communities. These range from camping facilities to homeless shelters, children's playgrounds to meditation spaces, refugee camps to eco-villages.

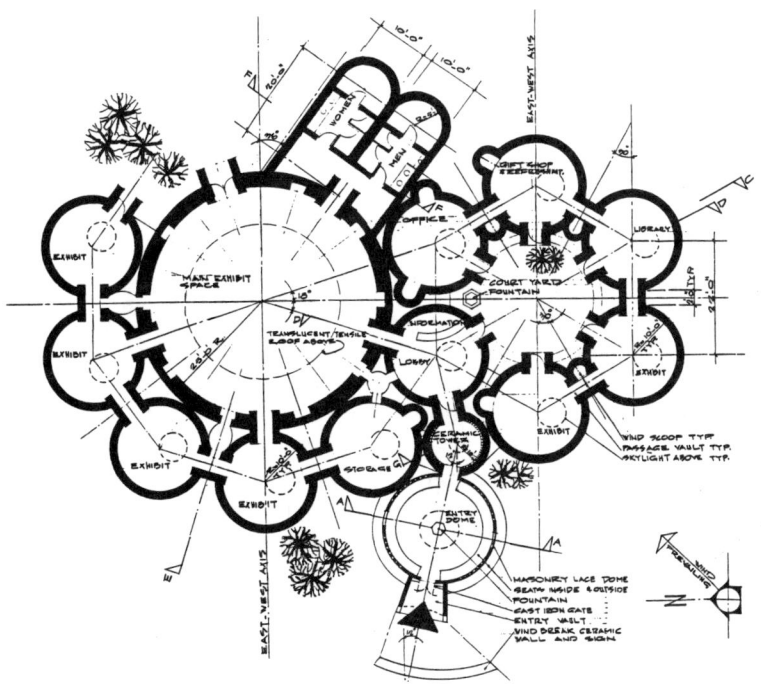

Left: Design for the Hesperia Museum.

Below: Natural clustering of the honeycomb in sevens.

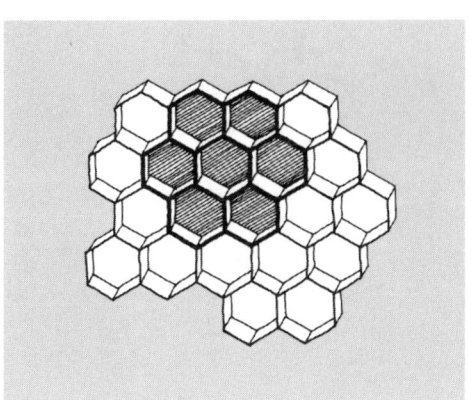

SUPERADOBE - SANDBAG SHELTER

Examples of a single wall pattern becoming the spine of a community.

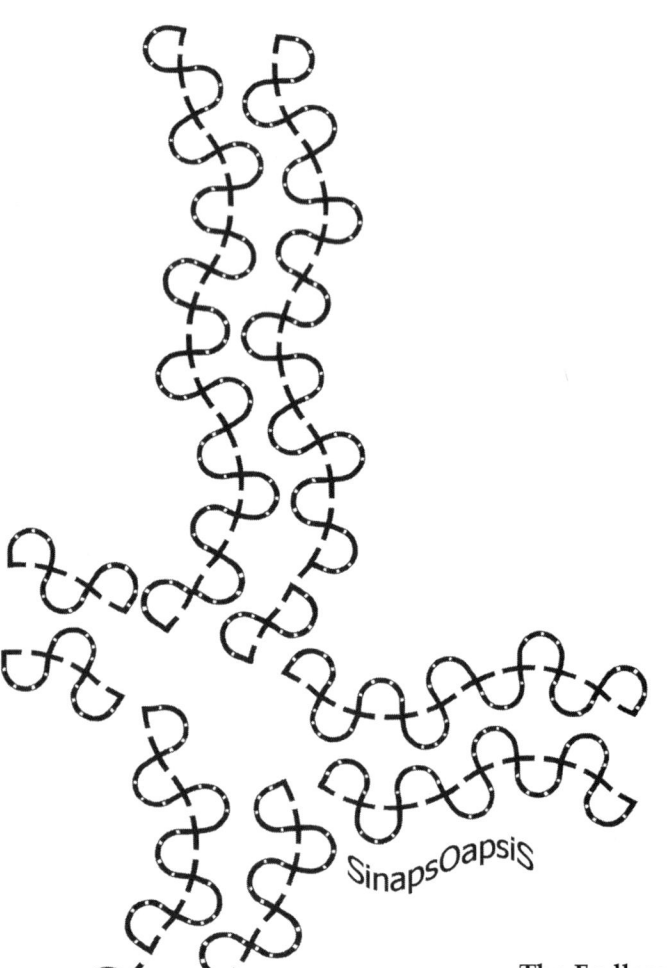

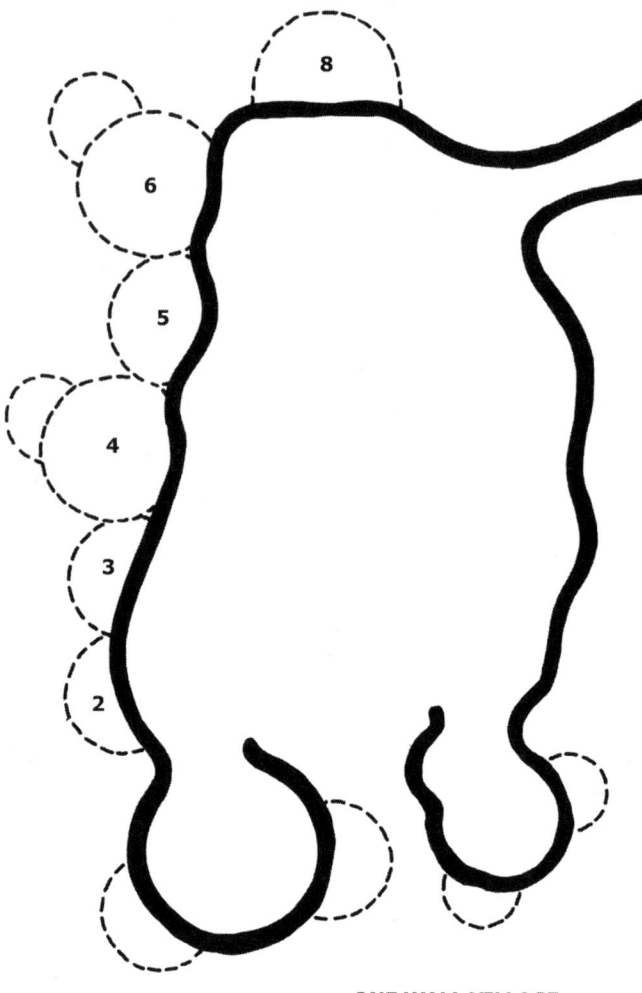

ONE WALL VILLAGE

The Endless Wall

The wall is a timeless principle that separates and unites, bringing together shade and sun, and protecting from adverse elements. Sharing common walls and creating courtyards can generate a neighborhood. A single wall can become the basis for a village or community as it follows the contour of the landscape, knitting together a fabric of apses, domes and vaults.

For the homeless, in congested cities, a thin strip of land could be dedicated at the edge of a park, road, river, or easement, to build a Superadobe wall, and from this wall apses/pouches can project as safe sleeping niches called "bed-wombs". For farmworkers, these walls and niches can surround the edge of fields, protecting the open land. The prototypes for this design are called "kangaroo pouches". When these are repeated in series they are given the name "Sinapsoapsis" or caterpillar. The continuous loops created by a single wall design concept can be used for mass housing in emergency situations. The system may utilize both manual and mechanized construction.

Above: A wall of domes, apses/pouches, seating and planters.

Right: The continuous coils of a Superadobe wall.

Left: Sitting in the shade of the wall at the entrance to a pouch.

SUPERADOBE - SANDBAG SHELTER

Right: Festive activities in the courtyard.

Above: Courtyard formed by shelter variations.

The Courtyard

When many emergency shelters cluster around a courtyard, a focus for community life is created, enhanced by steps, seats, planters, nooks, a fountain, and a dance floor or children's playground. As in a community of human beings, the shelters variations create unity from multiplicity.

The natural clustering of circles into seven, and their arrangement around community spaces can be either geometric or organic, depending on the terrain and the culture.

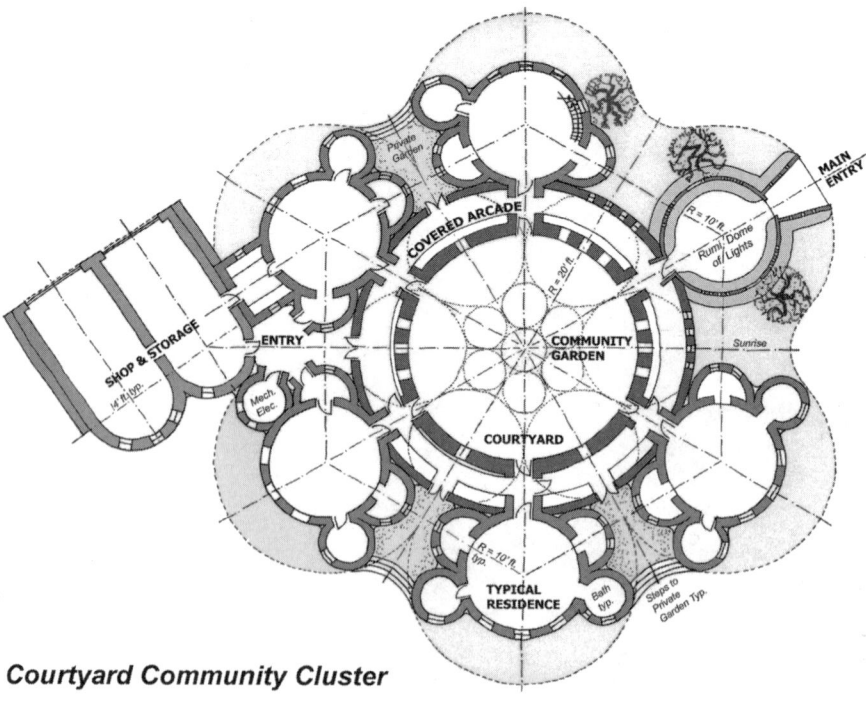

Courtyard Community Cluster

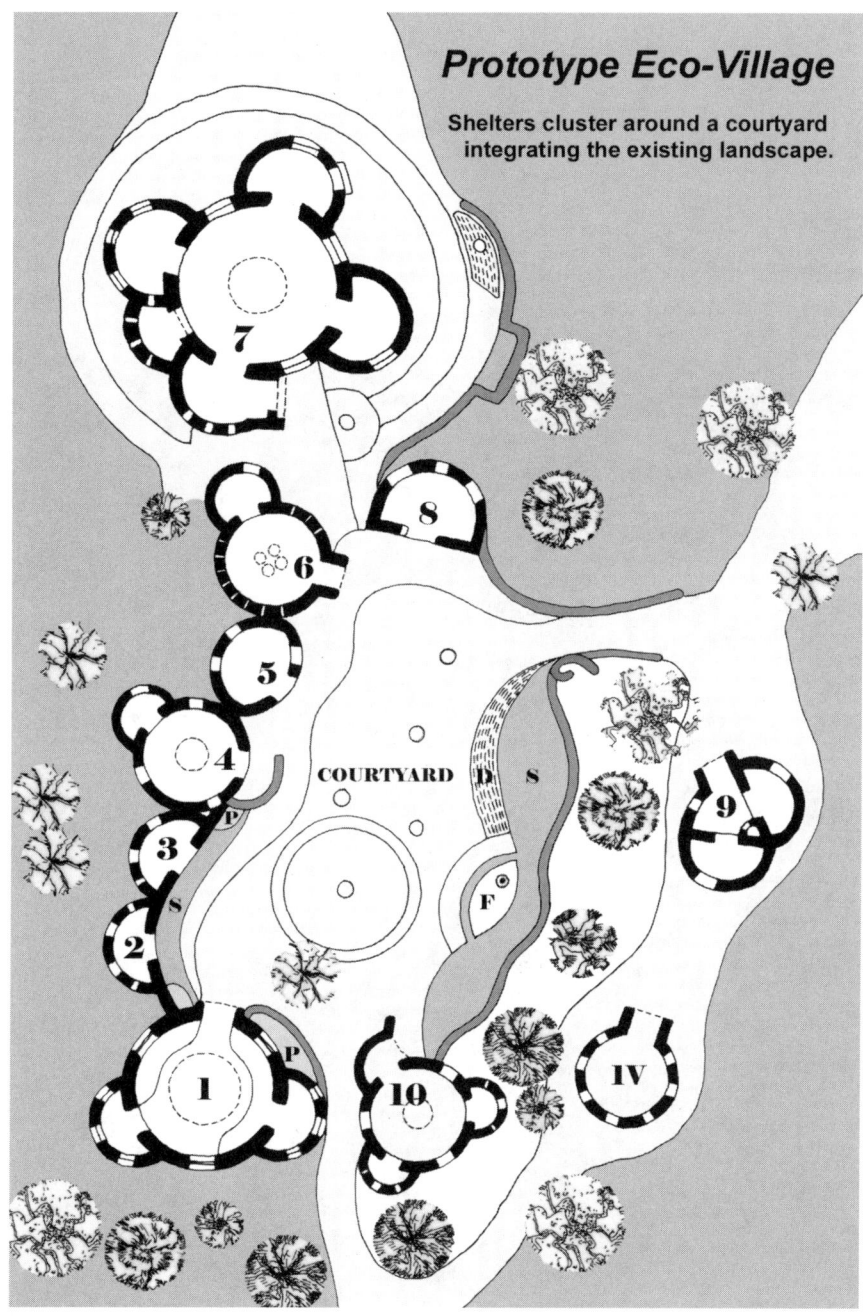

Prototype Eco-Village

Shelters cluster around a courtyard integrating the existing landscape.

LEGEND FOR MODELS:

1. Roofless Dome
2. & 3. Kangaroo Pouches
4. Dome of Potteries
5. Koala Dome
6. Holey Dome
7. Eco-Dome - Moon Cocoon Model
8. Sinapsoapsis - Caterpillar segment
9. Homeless Deluxe
10. Seashell Dome
IV. Compass Demonstration
F. Fountain
S. Seating
P. Planter
D. Drainage

Note: Some elements of the following shelter variations can only be achieved with stabilized earth, and should not be attempted with raw earth. The builder should work with small models, letting gravity and the earth's angle of repose dictate what can safely be built using the available earthen material.

Below: Courtyard view of shelter variations and fountain.

The Village Courtyard

The numbers on the Eco-Village plan above represent shelter variations, with names such as "Homeless Deluxe" and "Kangaroo Pouches", built around a courtyard. These demonstrate sizes for one person, two people, or up to a family size. The following series of photos describe the construction of most of the above shelter variations, teaching new techniques and designs which can be adapted to the various climates and needs of people worldwide.

SUPERADOBE - SANDBAG SHELTER

Top: Back view of a wall of shelters.

Above: Clustered shelters around a courtyard.

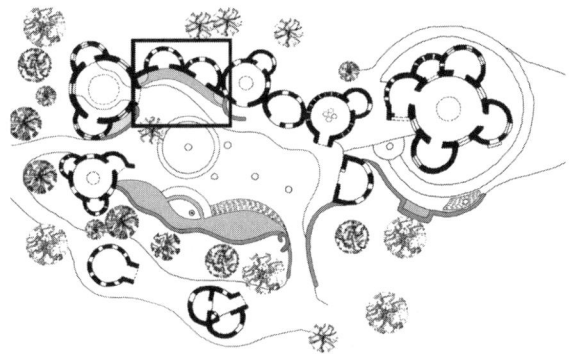

MODELS 2 & 3
POUCHES

SUPERADOBE - SANDBAG SHELTER 95

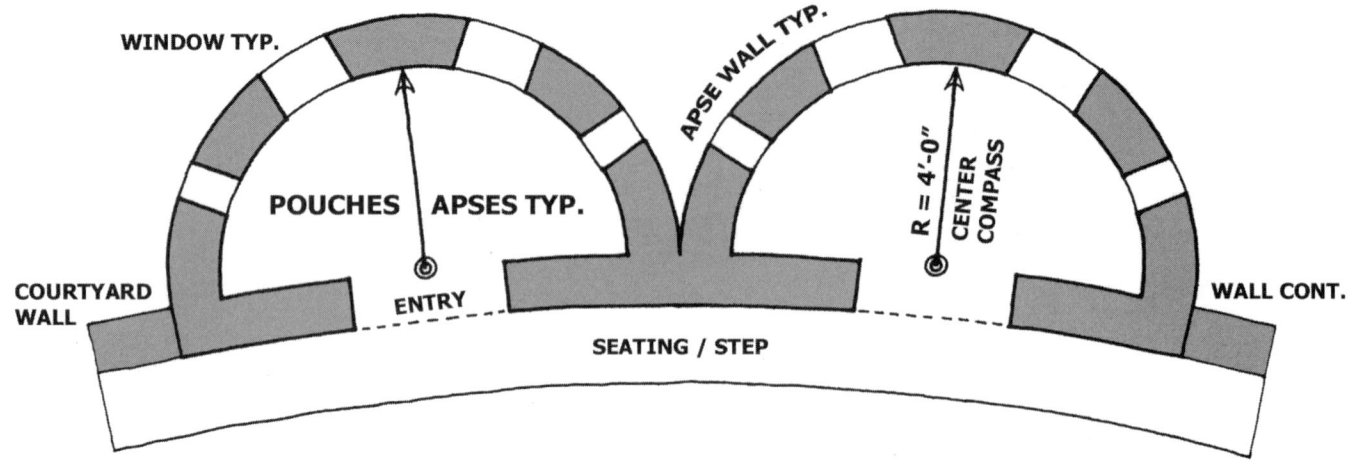

Above: Plan of the Pouches.

MODELS 2 & 3: Pouches

The "Kangaroo Pouches" are apses attached to a wall such as the courtyard wall. These two pouches are each eight feet in diameter and four feet deep. They are attached to the courtyard wall by the barbed wire and also by the bags which overlap.

Above and right: The foundation trench for the pouches is dug out next to the courtyard wall, levelled and then tamped. The bag is measured in place to fit the curved wall of a pouch.

The pouches and other domed shelters were built using the earth dug out from the courtyard, which becomes one to two feet lower than the interior floor of the pouches. Here, the courtyard wall was first built up as a small retaining wall, and then the pouches' foundation trench was dug out and compacted. After cutting the right length of long bag, the foundation row was filled and laid into the trench. Because the pouches are small (less than 10 ft. diameter), only one Superadobe row is needed below grade, and one strand of barbed wire per row. The barbed wire is placed along the center of the bags and overlapped into the courtyard wall by at least two feet.

A compass must be set up for each pouch to start building the curved form (see Appendix IV, Compass). The "center compass" is close to the courtyard wall, as shown on the plan.

Openings in the Pouches

For the basic emergency shelter you saw how to make arched openings and use pipes for windows. In the next shelter variations you will see several different methods used to create the windows by packing the bags over temporary "forms" of wood or metal.

Here, the larger arch forms for the pouch/apse entry are thin because the courtyard wall is vertical, whereas the small window forms are deep to accomodate the curving wall.

Above: The barbed wire connects the foundation bags of the niches with the continuous courtyard wall. Here, bricks hold the barbed wire down temporarily.

Below: Pouches with forms for windows and doors.

SUPERADOBE - SANDBAG SHELTER

Above and right: The pouches/apses at mid-height.

Below: The bags step in a little more at every row over the wood form to create the pouch/apse entry arch.

Arched openings which are tall and pointed like the pouch entryways can be built with unstabilized, raw earth, provided that the bags stop when they reach the wooden form on either side of the opening, and are neatly folded closed and compacted. The arched openings in the curved back wall, extend the long bag over the form, and therefore must be constructed with stabilized earth. The windows curve along with the curved wall both horizontally and vertically, since the pouch is an apse (section of a dome).

As you can see from the design of the first two pouches, they can be repeated almost endlessly along the wall which can be curved or straight. The pouches may also be partly set into a hillside, like small dams, turning the whole design into a stepped or "terraced" land-

Left and above: The pouches are near completion at the top. The rows get smaller and thinner, until the last small bag closes the top. A small round pipe opening helps with ventilation of hot air at the top of the pouch.

Below: A small window in the back wall of the pouches built with stabilized earth.

scape, growing from the earth.

The pouch entrance is easier to construct than a dome entrance, because it is a single arched opening in a vertical wall and therefore does not need an eyebrow or entry vault. In an emergency where materials are not available and speed is essential, pouches can be built using cut out entrances and pipe windows alone.

A mechanical mixer may be placed in the courtyard, and as near to the structure as possible, to mix the earth as it is dug out.

The barbed wire is the main element connecting the straight wall and the curved walls. However, in this example, the rows of bags are also overlapped at the corners. These sleeping pouches/apses are closed all the way to the top and there is no skylight.

SUPERADOBE - SANDBAG SHELTER

Above: The back of the pouches during construction. The mixer, cement, and water are close to the raw earth pile. The rolls of bags and barbed wire are on posts so that they can unroll quickly, and all the needed tools are as close to the structure as possible to minimize transporting and handling.

The builder works mainly from the outside to lay the rows of bags which are used as natural steps to climb up and down during construction, without needing a ladder or scaffold.

Scratch Coat - Rough Plastering

A scratch coat of stabilized earth plastering may start while the bags are being laid, and should be finished soon after the pouches are built to protect the fabric materials from disintegrating in the sun. It is best to use a native plaster mixture, tried and tested by local tradition, and

Right: A scratch coat of plaster is started during construction.

Above: More shelters are started next to the pouches, so that construction and finishing works can be completed together.

suited to the local climate. Here, the stabilized earth plaster has first been applied to the south side where the sun's rays are strongest. After thoroughly soaking the wall with water, a scratch coat is applied over the wall surface using a trowel, or by hand if the builder wears gloves. The surface of the scratch coat must remain rough, to allow the next plaster coat to bond well to it.

Left: Plastering is continued over the whole structure by hand or with a trowel.

SUPERADOBE - SANDBAG SHELTER

Above and right: The "scratch" and "brown" coats of plaster covering the bags.

Right: The seating is started with a long bag over arched forms.

Below: Shaping the planter and seating area. Here, the bags are laid on bricks as part of the decorative courtyard floor.

Seating

Seating can be created using a single long line of Superadobe over several arch forms. Any left over rubble (dry stabilized earth, bags, barbed wire, rocks etc.) can be mixed with a cementitious slurry and used to fill the gap between the long Superadobe and the pouches. This seat is about eighteen inches high and two feet deep.

The seat is then capped with fired brick pavers pressed into a two-inch thick stabilized earth surface. The surface is given a gentle slope to allow rainwater to shed onto the courtyard floor away from the

Left: Bags are laid over the forms to make a continuous arched seating. Behind them will be filled with earth and rubble.

shelters. To create a durable finish for the seat, a steel trowel is used to press (almost polish) the stabilized earth as it sets to leather hard. Finally, it can be gently brushed with a moist brush for a finer appearance and finished with a clear waterseal.

Below: Brushing the leather hard finish plaster coat, and sealing the surface after it has dried.

SUPERADOBE - SANDBAG SHELTER

Right: Defining lines are covered with white or colored stucco.

Stucco "Paint"

For the pouches, the rough stabilized earth plaster was smoothed by hand to make a natural finish, and the arches were highlighted with a white line. This was done with a commercial stucco mixed with water to a milky consistency (lime, sand, and cement) and then brushed onto the plaster. To preserve the seating finish, a layer of loose earth was temporarily spread over it.

The white line emphasizes the boundaries and openings of the organic geometrical shape of the structure, in the simplest manner, while providing opportunity for artistic or indigenous decorations.

Left: Finished pouches.

The flat wall on the southern face of the pouches absorbs and reflects the winter sun. Thus the combination of seating and the south facing wall create a warm outdoor social environment. The human voice resonates from the concave curve of the earth walls.

Below: A social gathering benefits from the radiating walls in the winter sun.

SUPERADOBE - SANDBAG SHELTER

Pouch, seating, planter and steps anticipate the human scale.

MODEL 4
POTTERY DOME

SUPERADOBE - SANDBAG SHELTER

Above: More domes continue after the pouches. The series of domes 4, 5, 6 & 8 were built at almost the same time.

Right (above): Dome of Potteries plan. The central compass is for the 9 ft. interior diameter dome, and the side compasses are for the apses/niches.

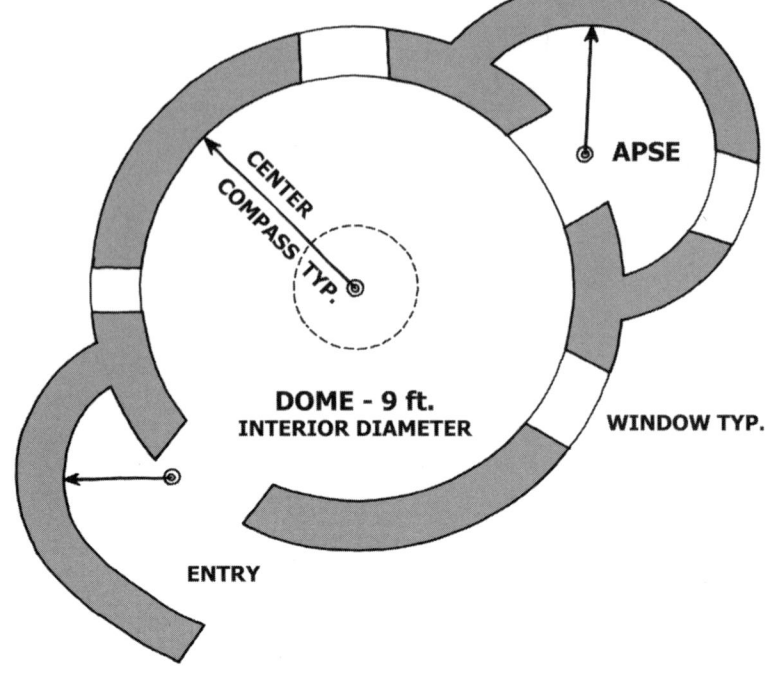

MODEL 4: Pottery Dome

The "Pottery Dome" is a family sized dome and apse of 9 ft. interior diameter dome built like the first emergency shelter, but using stabilized earth, with an apse/niche and an entrance protected by a curved wall. The size of this dome is large enough for a family's temporary shelter. Since it is larger and taller than the pouches, a full sized entrance arch is built over a wooden arch form.

Right: The wooden form for the entrance is supported on wedges. The bags are corbelled (stepped in) to make an arch.

Above: The construction crew can go from structure to stucture, and build several shelters at the same time.

Arch Varieties

Here you can see several different window forms; pipes up to 12 inch diameter and wooden arched forms for semi-circular and pointed arches. The bags can be laid in different ways over them.

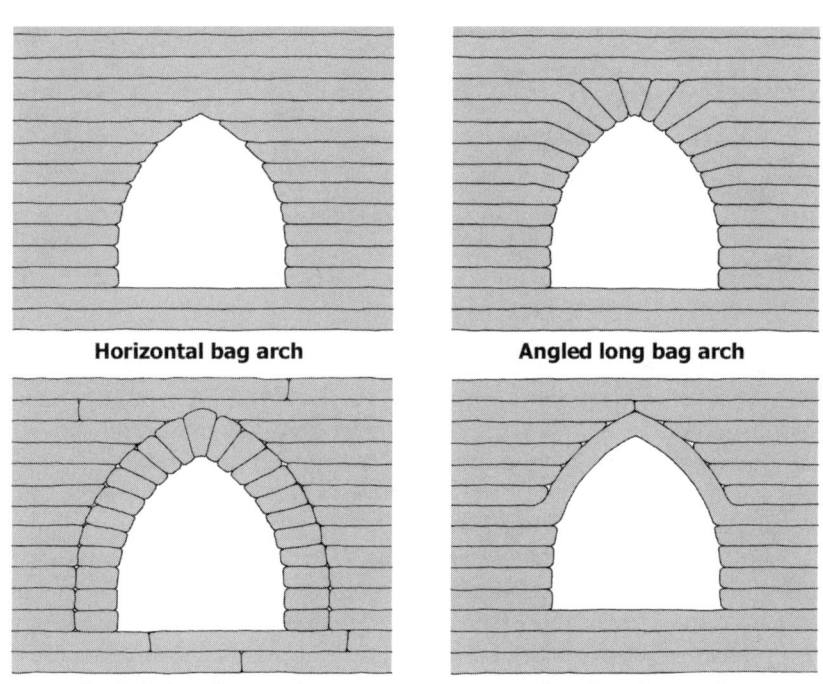

Left: Several different ways in which long or short bags can be placed to build a pointed arch.

SUPERADOBE - SANDBAG SHELTER

109

Right: Several different arch forms are used to build Models 4,5,6,and 8 , which are constructed at the same time.

    The forms for doors and windows are raised up on small shims/wedges of wood or pieces of brick. When the arch is completed these wedges are removed, allowing the form to drop down and be pulled out. Each form is deep enough to support the curve of the dome as it goes up over the window. The higher the dome, the greater curve, and the deeper the form will need to be.
    Window and door arches can be built by continuing the Superadobe rows up to the form and stopping, until the last row goes over the top of the form. Or, one long bag can become an arch over the form. Because these domes are small and built with stabilized earth, buttress walls were not needed for the doorway.

Above: Forming two types of arched entry; one long bag becomes an arch, or the bags are corbelled (stepped in) and stop at the form. Above the door arch is an upper level vent.

A form is almost always temporary and reusable, but sometimes it can be left in place to become the window or door casement. Many times the round pipes or steel forms are left in place, but the wooden forms are usually removed and reused. When the form is left in place it can create both the "eyebrow" and rough frame.

When there is no entry vault to protect the door opening, a cloth, plastic, or carpet covering can be attached into the dome wall and draped down to deflect wind and rain as described in Model 8. Small eyebrows over the windows could be shaped later if needed with stabilized earth plaster.

Upper Level Vents

At the top of most of these small shelters a pipe or two helps to ventilate by drawing off the hot air which rises to the top of the dome in summer. Since earth buildings have massive walls creating thermal mass, their interior temperature stays fairly constant and changes slowly. The domed shape also reduces the heating effect of the summer sun by creating sun and shade zones throughout the day.

SUPERADOBE - SANDBAG SHELTER

Above: "Now the time has come to create a new scale in the ceramic world, to walk out from the womb of a pot to the space of a room." quoted from *Racing Alone*.

Right: Interior coils of the dome, and the vertical pipe compass.

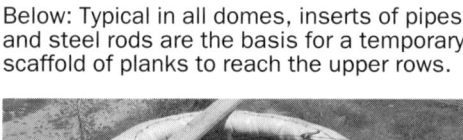

Below: Typical in all domes, inserts of pipes and steel rods are the basis for a temporary scaffold of planks to reach the upper rows.

Coiling Superadobe

The continuous Superadobe coils throughout the upper part of the dome demonstrate the flexiblity of the Superadobe system. The name "Pottery Dome" is a reminder of how closely the dome construction resembles the traditional coil pot. In future applications these structures may be generated mechanically using the principle of the potters wheel. The motion of the turning wheel and centrifugal force which create a pot translate into the turning compass and gravity's force for creating a dome. These make the natural symmetry of compression shell structures.

Coils of bags are continued in rows up to the top of the dome, measured and placed using a compass. Bag material lengths vary from about 12 to 30 feet long. The top of the windows especially, are carefully measured with the compass to form them along the dome's curvature. Just as a potter presses each coil while it is still soft to mold it to the adjacent coil, so the Superadobe builder tamps each row which hugs the previous rows, gripping the barbed wire

The vertical pole compass in the photo was made from a long steel pipe secured in the dome center, and was built to test this compass technique. Attached to the vertical pipe, a horizontal boom arm (a pipe with a clamp and marker) turned in a circle to measure the rows of bags. At every row, the clamp was raised up on the boom arm, and the horizontal marker was adjusted. It provided accurate measurements at the top of the dome. However, it is essential to use strong pipes which do not deform. This type of compass uses a different method to achieve the same domed shape as the two chain compass; just as arches can be made in several ways, so can domes.

Above: Exterior coils of the dome, and the vertical pipe in the center.

Left: Section of dome with pipe compass.

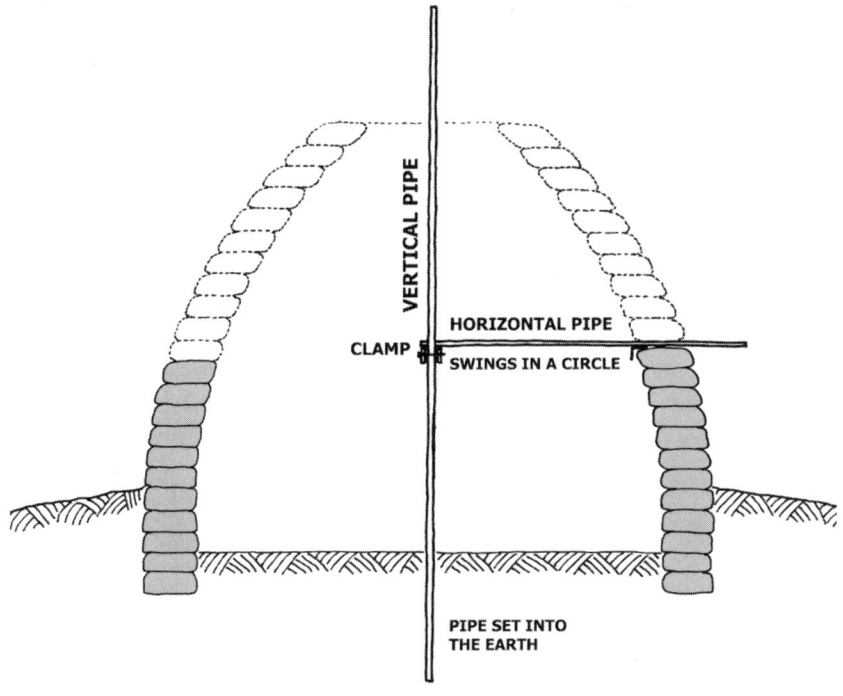

Below: A temporary scaffold of planks tied onto steel bars.

Scaffold

A small shelter which is set a few feet down into the earth can be built without extra hoists or scaffolding; standing on the natural grade outside is all that is needed to reach the top. However, when the dome is 10 ft. diameter or larger, the builders will need some steps or scaffolds on which to stand. By sandwiching short steel rods (rebars) or pipes between the coils and tying planks securely onto these, a temporary scaffold can be made which is both versatile and strong. When the dome is completed, the scaffold is usually dismantled and the plaster finishes are applied to the exterior.

SUPERADOBE - SANDBAG SHELTER

Right: After the bags are completed, the exterior plaster can be applied, as it was to the pouches, by hand or with a trowel.

The first plaster layers of "scratch coat" and "brown coat" can be applied directly over the bag material because the rounded shape of the coils provides a "key" which locks the plaster onto the wall. Additionally, the curving form of the dome gives the plaster layer some qualities of a thin shell structure. Both scratch and brown coats should have a rough surface for each layer to adhere to the next.

Right: View of the plastered Pottery Dome between the Kangaroo Pouches and the next shelter, the Koala Pouch. The finish plaster may be rough or smooth depending on the desired appearance.

MODEL 5
KOALA POUCH

SUPERADOBE - SANDBAG SHELTER   115

Right: Plan of the Koala Pouch.

Below: Geometry of the Koala Pouch.

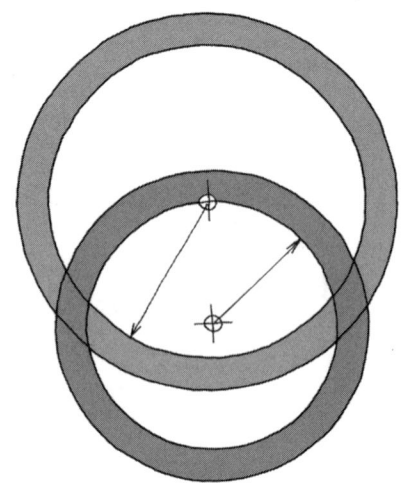

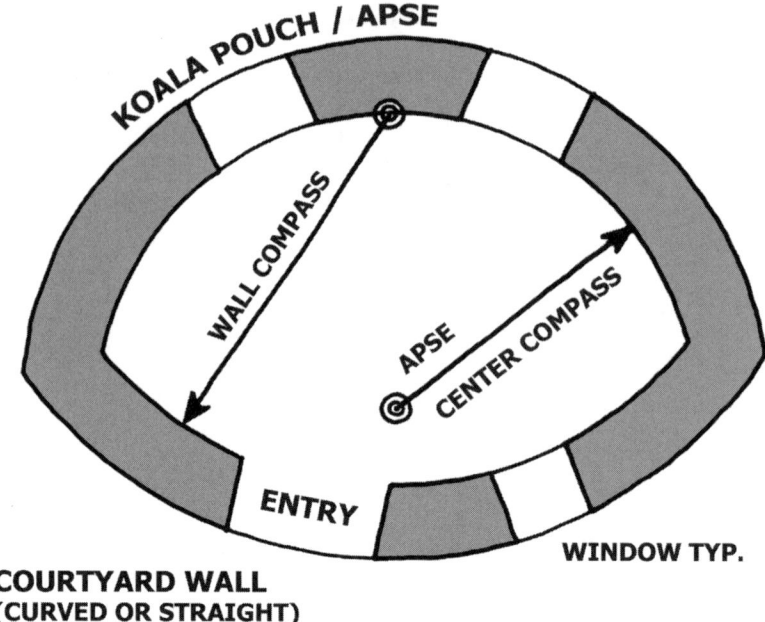

Below: Shelters 4, 5, 6, & 7, "Dome of Potteries", "Koala Pouch", "Holey Dome", and "Sinaposapsis", are built at the same time.

MODEL 5: Koala Pouch

The next shelter is a small one or two person shelter called "Koala Pouch" after its finished look, with a large white ring around the window resembling a Koala bear's face. It is the right size for a grown person to lie down inside.

Geometrically this domed shelter is made by overlapping two circles, like a stretched pouch. Conceptually it shows that not all domed shelters need to be circular.

Interlocking Corners

The curves of the front and back walls meet as a corner, and good construction of any corner means using the barbed wire and interlocking the bags with neat work. The barbed wire provides multiple tensile elements to resist cracking and helps to create safe structures which can withstand the forces of nature. When a stronger, firm connection is needed, the bag ends are interlocked as well.

In square or rectangular buildings the corners are the areas of high stress when an earthquake shakes them. It is usually the corners of a cube which crack.

When the Superadobe bags end on a corner or free standing wall, a brick can be used to shape them, for neat and well-filled ends, before tucking under to close, and tamping. When the bag starts on the corner, a metal sheet can assist the builder to position it accurately (see Model 8 "Sinapsoapsis" for details). Twisting the ends, as shown in the photo, can also be used to turn the corners like a long sausage, to

Left: Neatly overlapped ends are connected by barbed wire. A single bag makes the entry arch.

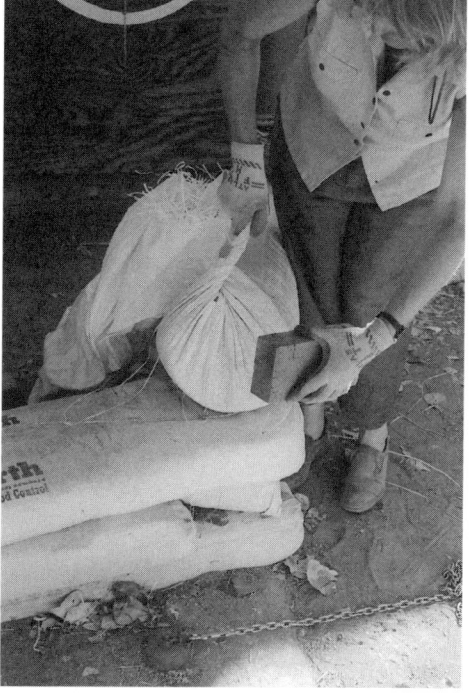

Below: To get a neat end which does not droop, twist the bag and compact with a brick.

continue filling without cutting the bag.

In this architecture, when domes, apses and vaults are clustered together, it is important to understand when the corner connections must be firm and when they must be flexible.

Long Bag Arch

The wooden arch form for the entry is placed on wedges before the long bag is built over it. One bag is placed either side of the form to hold it in place, then one long bag makes the arch.

SUPERADOBE - SANDBAG SHELTER

Right: Continuous long bag arch for the entry and a 12 inch diameter pipe window.

Below: Crumpling/corrugating the bag onto a tube and pulling the end through.

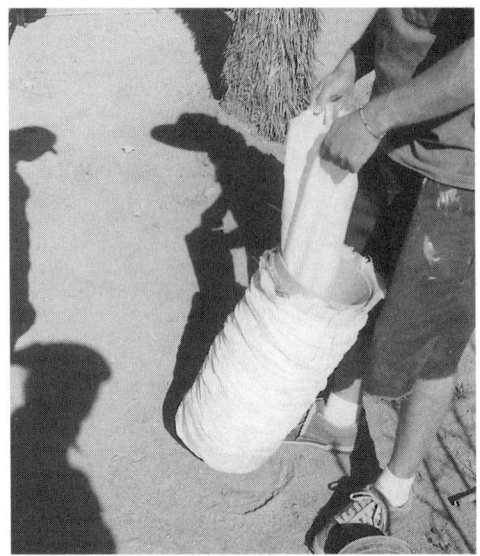

Because the entrance wall of the pouch is almost vertical, the bag which makes the entry arch goes straight up over the form and down the other side. It can become a short or tall door. No "eyebrow" or overhang will be needed to protect from rain, although some sculpting with the exterior plaster will create a drip line/edge.

Curves and Compass

The back wall curves to make the apse, while the entrance wall is slightly bowed. To achieve these curves each wall has its own compass set up, which may be pre-measured, or be a two chain compass system (see the Compass Appendix). Being a small structure, it can also be judged by eye without needing a compass. With or without a compass, the curves must be smooth and gradual.

Above: Filling the bag using the tube.

Right: About 20 - 30 feet ready on the tube.

Left: As the weight of earth pulls down on the bag, the top unfolds and more bag is released.

Below: The bag is fed through the tube, with one to three cans of earth mix at a time, depending on the builder's strength and preference.

Filling Bags with a Funnel Tube

The builder uses a lightweight funnel tube to fill long lengths of bag. A length of the bag material is pulled (crumpled/corrugated) onto the tube; about 20 - 30 feet can be pushed onto a three foot long tube. The last few feet of bag material is passed back through the middle of the tube, which stops the bag slipping off the tube when it is filled. The earth is poured in through the tube; the builder can control the supply of earth where he is placing it on the wall and unfold more bag as needed. This method is especially useful when the earth mix is nearby, and can eliminate the need for an extra builder.

Above: A small shelter's human capacity.

Below: The exterior plaster is pressed into the groove between the bags.

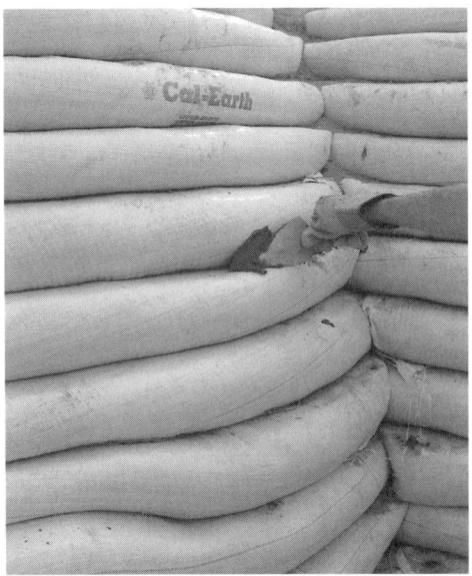

Stem Wall of Stabilized Earth

It is sometimes possible to use a stabilized earth foundation and stem wall without bags, as was done on domes number 4, 5, 6, & 8, whose walls start two feet below grade. The following system is only suitable with well stabilized earth or concrete.

First, the plan or "footprint" of the five domes was dug down to make a deep foundation trench. The trench was filled with stabilized earth, layered with barbed wire every vertical foot, and stopped two bags below grade. After that, barbed wire and the first two bags below grade were set onto this small stem wall, allowing the natural grade to buttress their connection.

Here, the entire courtyard is sunken down into the earth by two feet and the back of the domes stays below grade. In the photo above the stem wall is all below grade, but when the courtyard earth is dug out the front part will become exposed. This is acceptable only when the domes are small; for larger domes, the stem wall and two bag rows should remain below grade.

Notice also that the entryways do not have buttress walls because they are not needed.

Above: The stem wall makes a step up.

Flood Control Entry

This "Koala Pouch" and the next "Sinapsoapsis" shelter demonstrate a stepped up entry which can form the basis for communities in flood prone areas. The height of the entry can be decided by the flood levels. Streets can be designed to take away the flood waters, whilst stepping stones and islands enable pedestrians to cross. In high flood areas, the natural steps created by the bag rows, or specially created Superadobe steps and small bridges can be designed allowing people access to the entries by travelling above the plaza floor.

Smooth Plastering

The plastering for the Koala pouch continues with an exterior plaster of stabilized earth, as was done on previous shelters. Firstly, the scratch coat is applied directly over the bags and pressed well into the grooves. Next, a thick brown coat is applied over the scratch coat, sticking onto it. If needed, a water resistant layer using asphalt can be mopped or brushed over the bags after the grooves are filled, or asphalt emulsion can be mixed with the brown coat.

Below: A scratch coat of plaster follows the grooves, then a brown coat goes over it.

SUPERADOBE - SANDBAG SHELTER

Above: Coiled structures before plastering.

Right: Scratch plaster coat applied by hand.

Below: Using a steel trowel for the finish.

The stabilized earth plaster is a similar mixture to the stabilized earth inside the bags, for a similar rate of expansion and contraction. The plaster has a higher percentage of stabilizer (cement, lime, or other stabilizer) and is as wet as conventional stucco plaster. Before applying the plaster, the wall is thoroughly sprayed with water.

Before plastering it was easy to climb up and down the dome using the natural steps of the coils; now we need a ladder to reach the top. If no ladder is available, steps can be sculpted into the plaster. Short sticks or rods may also be sandwiched between the Superadobe coils during construction and left protruding from the wall surface. Many traditional earth buildings around the world have used such a built-in ladder for plastering and maintenance.

Finally, a finish coat of fine plaster made from screened earth mixed with a higher percentage of the stabilizer, is spread over the dome. Pressing hard with a steel trowel brings the finest earth particles and the cement/lime stabilizer to the surface, resulting in a denser, harder finish. It is important not to wet or wash the surface at this stage. When the surface has set to leather hard, gentle working with a damp sponge can blend a uniform smooth finish. A clear waterseal can be mixed with the finish plaster, or painted over it.

Above: The finish plaster coat is applied over the brown coat.

Below: Gently massaging in circles with a damp sponge.

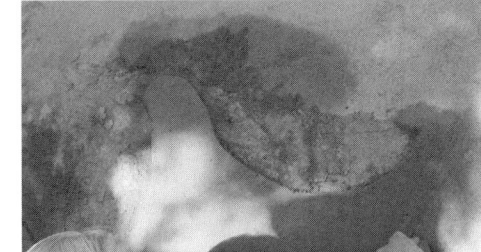

SUPERADOBE - SANDBAG SHELTER

Right: Sunlight and sunflowers - our favorite flowers for this architecture, express the organic forms and details of the human body. The lines of the plastered and painted shelters celebrate a theater for the human presence.

After the domes are smooth plastered, a layer of white stucco used like paint highlights their forms. Here you can see the entrance to the Koala Dome with its round, Koala-like eye, and the round window of the Dome of Potteries, with seating and a planter in front. Between the domes, white-painted steps invite one to climb up. The painted lines at the boundaries of top and bottom, steps, entrances and windows, speak the language of an architecture that could be adapted with different decorations in different cultures. Here, the simple white lines bring alive the intangible aspects of the courtyard at a human scale and nourish social life.

Left: The form and finish of the plastered and painted shelters, timeless forms awaiting a human drama.

Below: Social life at the scale of children.

SUPERADOBE - SANDBAG SHELTER   125

Above: Mother and child tamp the bags, demonstrating that a family can build their own shelter.

MODEL 6
HOLEY DOME

SUPERADOBE - SANDBAG SHELTER 127

Plan of the Holey Dome.

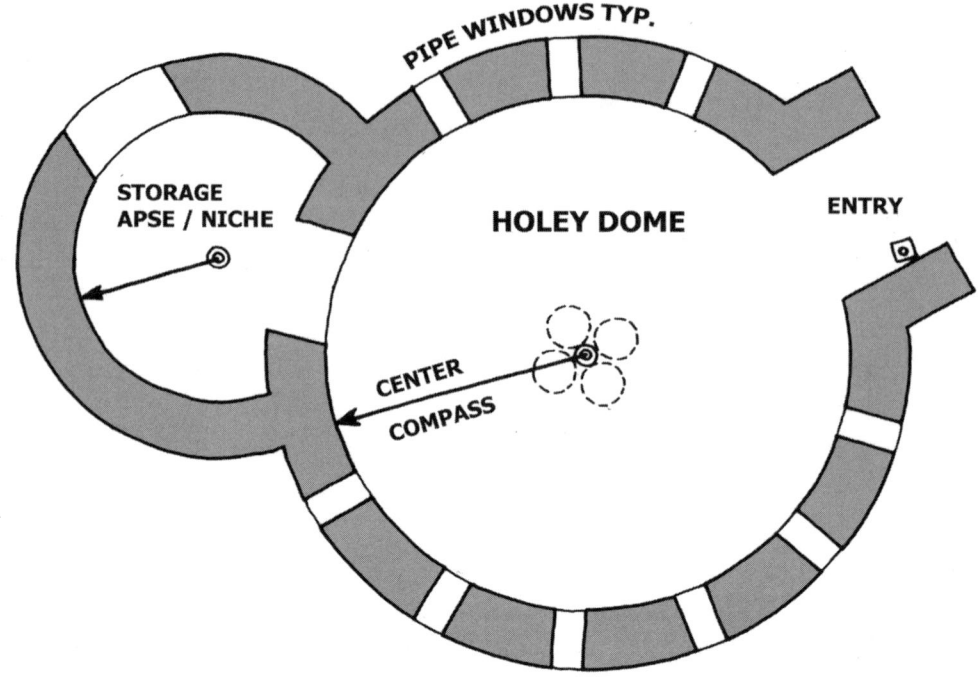

MODEL 6: Holey Dome

In humid climates or by the seashore continuous breezes are necessary through any building. With multiple pipe windows this circular dome creates comfort and beauty by the play of light and air. Additionally, pipe openings instead of windows can be a good protection in war zones. When pipes are placed at an angle they can deflect the path of bullets and shrapnel.

Right: Constructing a dome with many round openings using pipe sections.

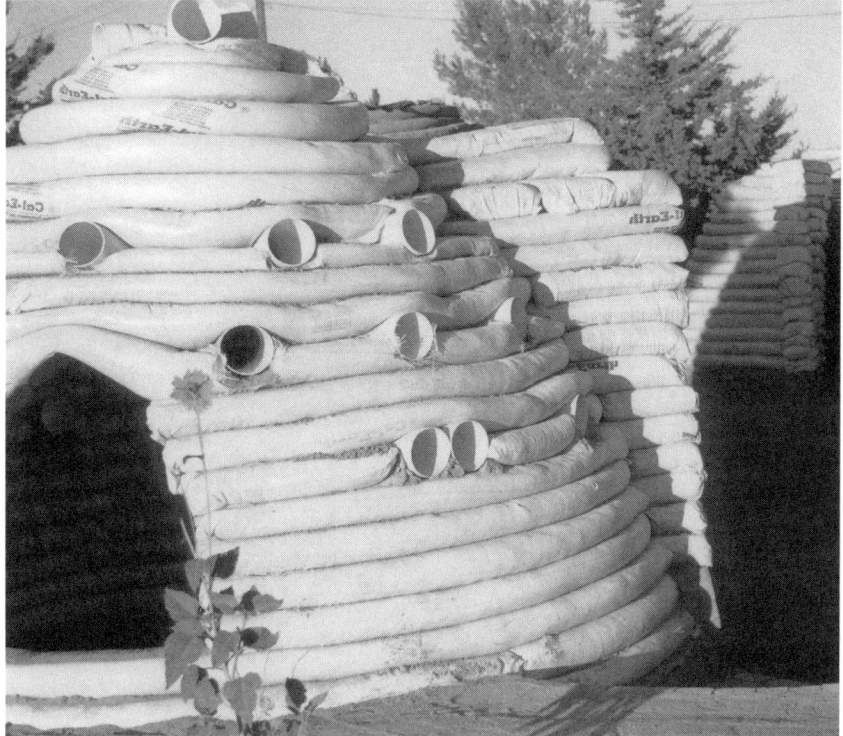

Left: Multiple pipes can be inserted every few rows for maximum ventilation.

Maximum Ventilation

For maximum light and ventilation thoughout the dome many pipes can be used because small round openings have minimum structural effect. The large opening above connects to a storage apse/niche.

In wet climates the foundation can be protected by a waterproof membrane (plastic sheeting) or easy drainage such as gravel and perforated pipes (french drain).

Above: Plastic sheeting below the foundation rows of bags protects from rising ground water.

Left: Interior view of round openings.

SUPERADOBE - SANDBAG SHELTER

Right: Cutting the channel for a pipe with a standard saw when the bag is still soft.

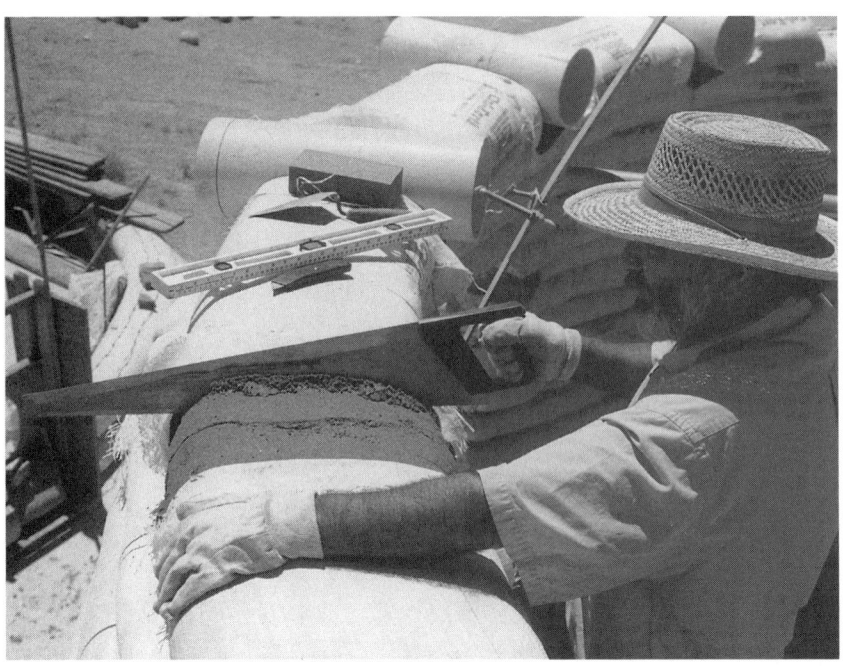

Pipe Window and Vent Setting

Earlier you have already seen how to insert pipe windows and vents by continuing the bag over the pipe and tamping it well. Since this Holey Dome has many round windows placed every third Superadobe row up to the top of the dome, you will now need to use the more precise technique of cutting the bag and scooping out the earth mix.

As you build higher up the dome, the weight of bags corbelling inwards will cause the pipes to slope inwards unless the slope is cut into the bag beforehand. By cutting and carving a sloped channel into

Right: Carving a slope and channel.

Left: Fitting the pipe which is several inches wider than the wall.

a freshly tamped bag you can keep all the pipes sloping outwards to drain the rainwater to the outside. This can only be done while the earth mix is still damp and soft. The photos show a step by step this technique for inserting the pipe windows.
1) First, the bag material is cut and peeled back. Don't cut it off since it can be folded back to protect the exposed earth fill later on.
2) Next, with a steel trowel or other tool, scrape out a sloped channel and fit the pipe into it.
3) Check that the pipe slopes towards the outside using a carpenters level, or test with water to see if the rainwater will run out.
4) After the pipes are set in place, the barbed wire is continued over them to connect the rows together. If many pipe windows are inserted, the barbed wire must be used above and below the pipes for a continous tension element gripping onto the bag surfaces.

Above: Connecting over the pipes with barbed wire; in a emergency shelter of less than 12 ft. diameter, a single wire will be enough.

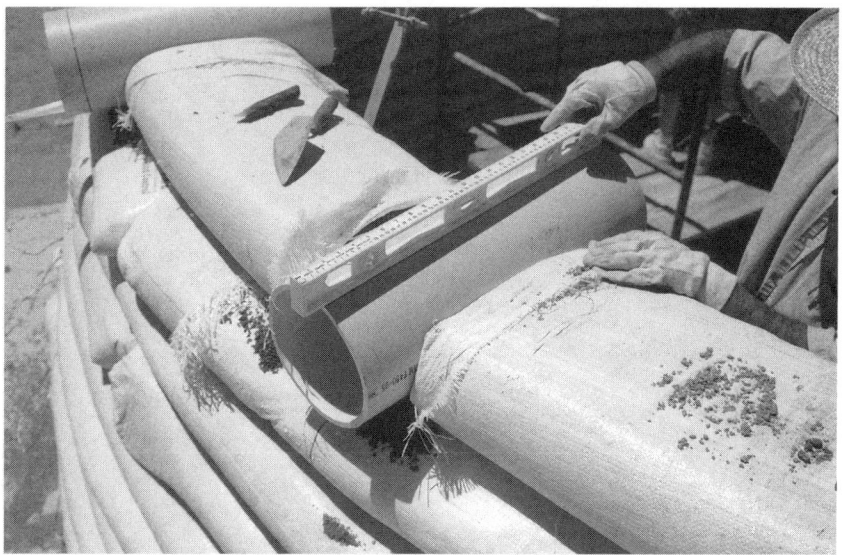

Left: Checking the slope.

SUPERADOBE - SANDBAG SHELTER

Right: Cans waiting to be used.

Below: Four 90 degree elbow pipes are tied together with wire to create the skylight and windscoop at the top of the dome.

At the top of the dome pre-filled cans can be collected before using them, or one person can throw up the can of earth and the second person fill the bag.

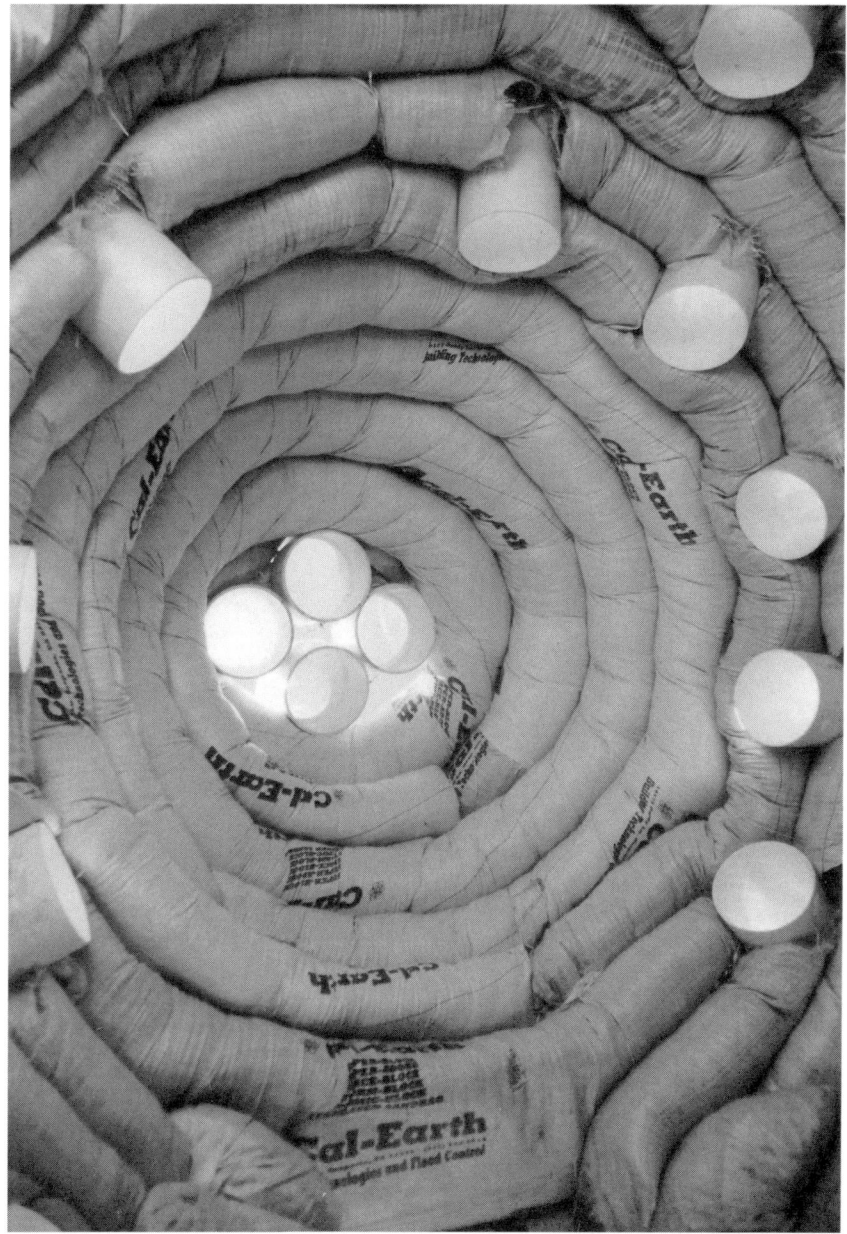

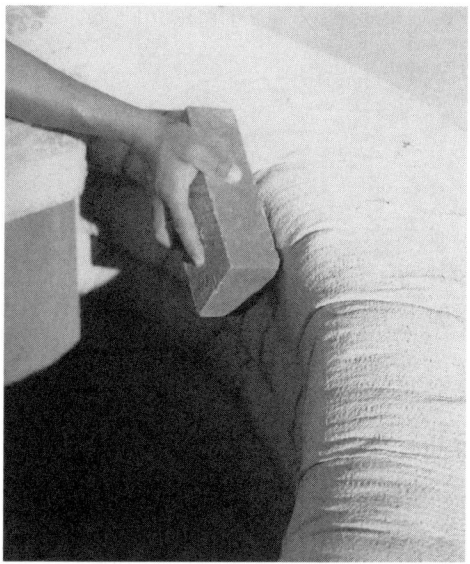

Above: Tamping the corbelled bags to shape the inner edge.

Left: Looking up into the top of the dome with the four-directional wind scoop pipes.

The Superadobe rings at the top of the dome make very tight curves to create smaller and smaller donut rings. If the builder immediately shapes the bag with a brick after each two cans of earth has been positioned according to the compass, these very tight curves can be achieved. It is possible for one person to build alone by holding the bag in one hand, and the brick in the other. It is good practice to shape the inner edge of the bag rings for a smoother curve to the dome.

At the top of the Holey Dome is an innovative skylight which doubles as a four-directional windscoop. It is made from four 90° degree bend drainage pipe elbows tied together, to catch cooling winds from all directions.

SUPERADOBE - SANDBAG SHELTER

Right: Finished Holey dome structure.

"Mineshaft" Entry

The entrance to the Holey Dome is not arched but is almost triangular with a short, flat lintel. As a very small entry it can be quicker to learn than an arch in an emergency situation. The entrance buttress walls are gradually stepped inwards until the distance between them is short enough to span. The lintel bag should be tightly packed with stabilized earth and then pinned closed before it is lowered across the entryway. Two or more strands of tensioned barbed wire stretch underneath each lintel bag, and a temporary support such as a piece of wood, should also be placed underneath to facilitate the tamping, which is necessary for the lintel's strength.

    This type of entrance resembles the old mineshafts, and like the mineshaft, it can make a long entryway but is limited in width. The Holey Dome has a small lintel about 18 inches span of stabilized earth. Any spans over 2 ft. should be reinforced with steel and buttressed. Unstabilized earth should not be used for lintels.

Left: Applying a layer of liquid asphalt over the upper curves of the dome.

Below: Placing the Reptiles directly onto the lower vertical wall of the dome.

Reptile Exterior

"Reptiles" are a cellular plaster that can be both the brown coat and finish coat. They solve the cracking problems of smooth plaster because they embrace the natural law of cracks. Hundreds of cracks between the reptiles enable the whole surface to expand and contract without breaking the surface.

The finished dome is first waterproofed with a layer of asphalt which is either painted or mopped over the bag material, or spread over a scratch coat of stabilized earth plaster. The first coat should be liquid enough to soak into the earthen material. The upper areas are waterproofed but the lower vertical walls these do not need waterproofing in drier climates.

Then the exterior is covered with Reptile stabilized earth plaster. The wall is soaked with water beforehand to help the Reptiles/mudballs to adhere. They will also adhere to each other, forming a domed shell of Reptile plaster.

Above and below: The Reptiles/mud-balls are playfully patted into shape, by throwing and catching them to eliminate air pockets.

Right: Placing the mud-balls in a reptile pattern on the wall.

Reptiles require less skill than smooth plaster. A handful of stabilized earth mix is made into a dense, smooth ball around the size of a tennis ball or larger. Stabilizing the earth with white cement and/or lime results in a natural earth tone color. Before placing the ball, dip your glove in water and moisten the surface for a more durable,

136   KHALILI - CAL-EARTH

Left: Placing the Reptile mud-balls on the wall. The balls are put in place, normally starting from the bottom and layering up to the top.

adhering Reptile ball. Then each one is dropped onto the dome wall so that it sticks. Throw too hard and you get a splat; not hard enough and it falls off. The Reptiles are overlapped like regular tiles, starting from the base and working upwards. Different colored balls can create patterns by adding natural pigments to the earth mix.

Left (below): Application of the Reptiles is therapeutical and invites community participation, including cihldren.

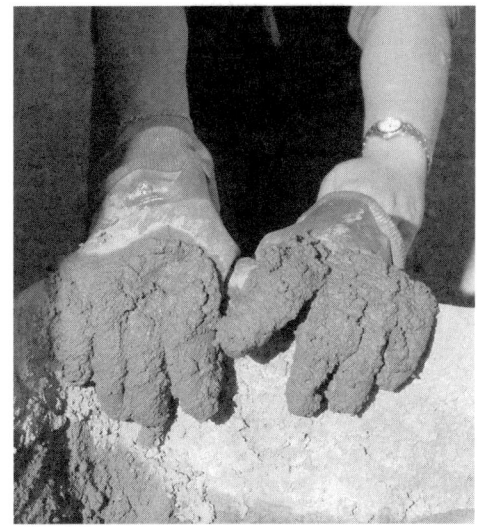

Below: Plastic gloves are used when cement or lime is part of the earth mix.

SUPERADOBE - SANDBAG SHELTER

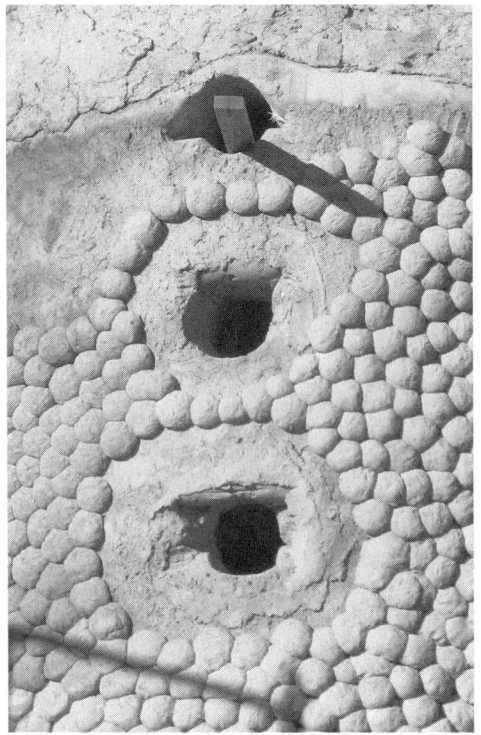

Above and right: Reptile exterior finish around many vents and pipe windows.

Below: Larger and smaller Reptiles are formed by larger and smaller hands.

Right: A finished pipe window. The glass pivots open and closed.

Later, the pipe openings can be upgraded to include a fixed pane of glass or an openable glass or plexiglass circle, like the pipe windows shown earlier.

Reptiles follow the natural law of cracks which says that every skin has cracks, from fish scales to the tree bark and reptile skins. Even the clay earth when it dries makes the same cracks. Reptiles solve the

cracking problems of smooth plaster because they embrace the natural law of cracks. Just like our own bodies and our flexible human skin which adds another crease when it is needed to stay flexible. Thousands of cracks mean that Reptile plaster can expand and contract depending on the curved surfaces underneath.

Their domed shape creates sun and shade zones within the

Left: Children compare hand sizes with the reptiles.

Above: Fish scales pattern.

Below: The natural cracks of clay earth.

Left: Turtle shell pattern.

SUPERADOBE - SANDBAG SHELTER   139

Above: Children experience climbing between the shelters, and touching the reptiles.

Right: A flower emerges from the reptile patterns, painted by an artist.

exterior skin, softening the sun's glare and creating a micro-environment like a reptile which relies on the sun to heat its bloodstream. Their rounded form invites touching and interacting with the surface.

We can look at the wall of Reptiles and find their distinct geometries within many random patterns. Following the geometry of seven circles, images which come to mind are sunflowers, waves, and the abstract colorful patterns of fish scales and reptile skins.

Above: Front view of the finished Holey Dome, with flat "Mineshaft" entry.

Left: Reptile detail with sun and shade.

SUPERADOBE - SANDBAG SHELTER 141

Right: Section through a dome with attachments for a tropical roofing.

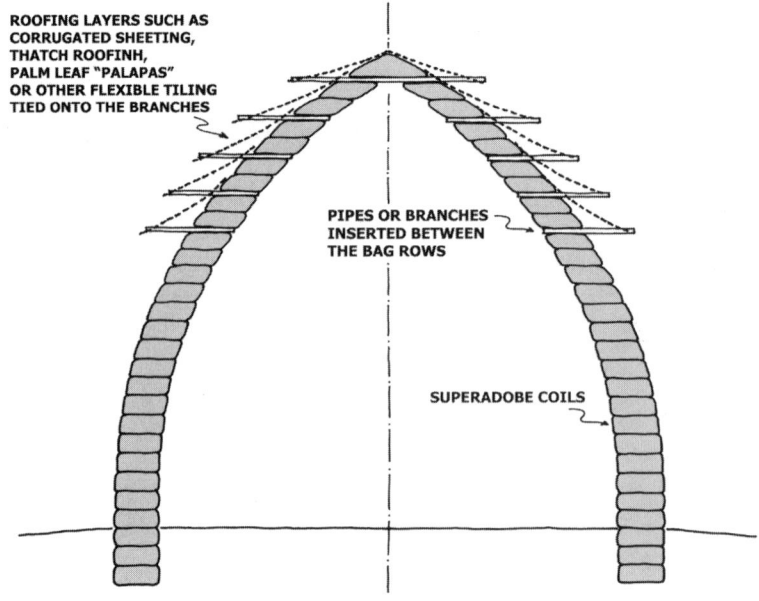

Below: Examples of traditional "Palapas" roofing for tropical climates.

Extra Roofing Layer (Porcupine-Style)

The Holey Dome model allows plenty of ventilation needed in hot, humid climates, but it will require additional roofing for tropical climates with high rainfall. Thus, the principle of overlapping Reptiles must evolve to an overlapping tile of water resistant material, which is flexible enough to accomodate the building's curved forms.

Traditional solutions of palm leaves (palapas) or corrugated sheet metal may be attached as overlapping elements, supported by the pipes or branches which were sandwiched between the bags as vents or scaffolding. Overlapping tiled elements may also be woven together and draped over the dome, like traditional armor or the scales of the armadillo. When the pipes or branches stick out porcupine-style they help to shed the rainwater away from the building.

Right: Smaller pipes, reinforcing steel, or branches are first used as a builders' scaffold, and later become elements for attaching a lightweight roofing.

MODEL 8
SINAPSOAPSIS
CATERPILLAR

SUPERADOBE - SANDBAG SHELTER

Right: Plan of the Sinapsoapsis shelter.

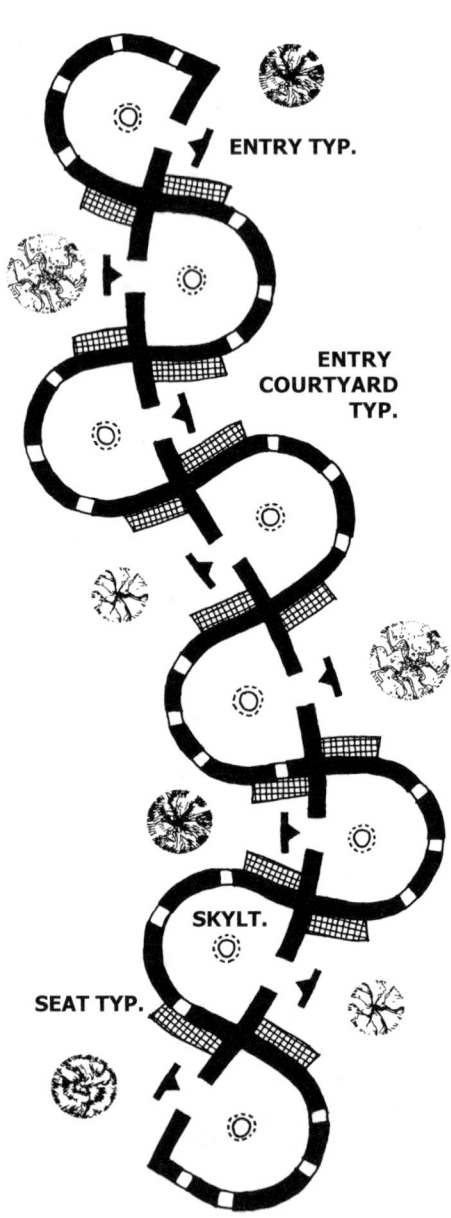

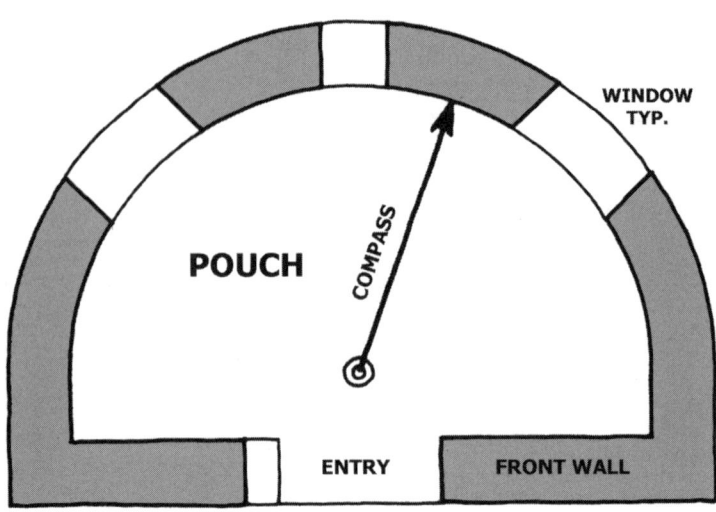

MODEL 8: Sinapsoapsis - Caterpillar.

The shelter at the end of the village plaza is a combination of a pouch and a dome. In this case, the straight front wall gradually merges into a dome towards the top, similar to a classical squinch dome.

Above: Multiple pouches make a Sinapsoapsis (Caterpillar), for large numbers of shelters by mechanized systems.

Right: The series of shelters along the courtyard wall making a partial Sinapsoapsis.

Left: Students are learning the squinch pouch (square becoming a circle at the top).

The Kangaroo Pouches, Koala Pouch and this squinch pouch are all examples of units which can be combined into a Sinapsoapsis. This last model No. 8 has more floor area because the compass center is set up to a foot inside the shelter and not close to the straight wall.

For large numbers of emergency shelters, long caterpillars of Sinapsoapsis construction can be mechanized using standard pumping and mixing equipment. Long Superadobe tubes can be pumped and coiled in place to quickly shelter many people. Streets, courtyards and clusters can be assembled from these long rows of pouches resulting in protected community spaces.

Left: A family, young and old, can build a shelter together. Concrete blocks or other wedges can the the base for a temporary arch form.

SUPERADOBE - SANDBAG SHELTER

Above: Sinapsoapsis exterior view from above. Note how the square corners at the bottom gradually become round coils at the top (a square changing into a circle, a dome).

Right: Interior view showing the lower layers making the corner, and the upper layers gradually curving into continuous coils to make a dome.

Below: The upper layers coil into a dome with a small skylight.

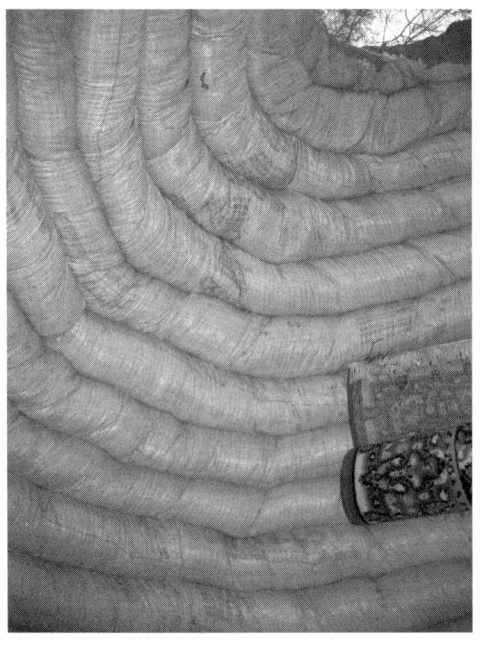

Merging Straight with Round

Traditional masonry techniques use "squinches" to put a round dome over a square room (*see Ceramic Houses and Earth Architecture*). Here, the continuous sandbags step in, or corbel, a little at every row which gradually merges the straight with the round.

In the beginning we can start building in the same way as the Kangaroo Pouches. A piece of straight wall in front is connected to a curved wall around the back, with the barbed wires and bags overlapped at the corners. The vertical, straight wall in front makes it easier to include an entrance.

Higher up the shelter, the curved wall bags around the back corbel inwards to make a dome, while the straight wall bags in front gradually curve backwards, until the front and back bags join to complete a circular coil. The skylight opening at the top is optional.

Straight Wall Entrance

The straight wall simplifies the entrance/door opening. A door, heavy cloth or carpet can be hung on the inside or outside of the opening. Doors have traditionally been made from layers of cloth, animal skins, or carpets by many indigenous peoples. Today, fabrics and plastics with zippers and velcro are used as doors into tents and warehouses. If needed, a lintel or "eyebrow" can be added by sandwiching suitable materials between the bags above the door opening.

Here the entry and interior floor are raised above the courtyard level to protect from rain, floodwaters or uninvited animals. The entry design can be straight or create a shelf for a lantern or flower pot.

Above: Carpets can cover the floor and the door opening; here one is attached on the inside.

Left: Exterior view showing the straight wall in front merging with the round dome.

Below: A flexible fabric door such as a carpet can also cover the door opening from outside.

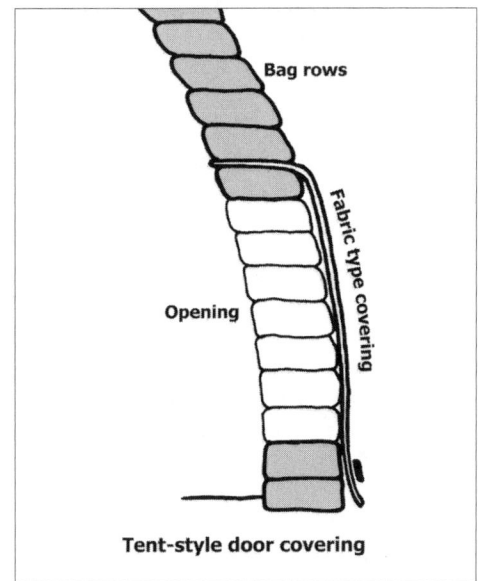

SUPERADOBE - SANDBAG SHELTER

Above: The entrance to the finished shelter.

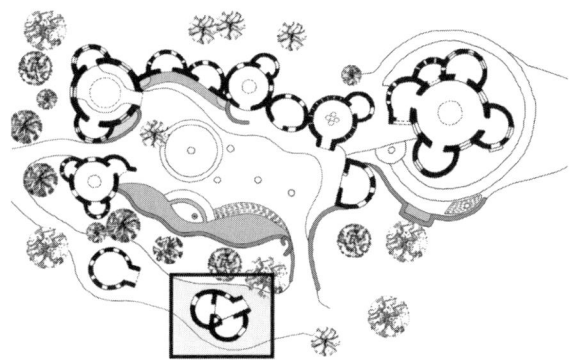

MODEL 9
HOMELESS DELUXE

SUPERADOBE - SANDBAG SHELTER

Right: Plan of the Homeless Deluxe.

Below: Geometry of the Homeless Deluxe.

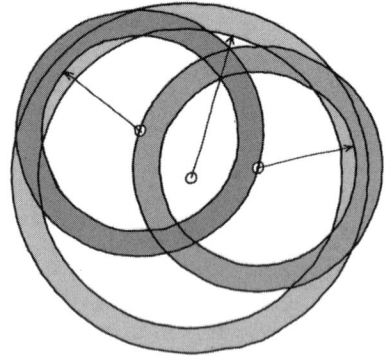

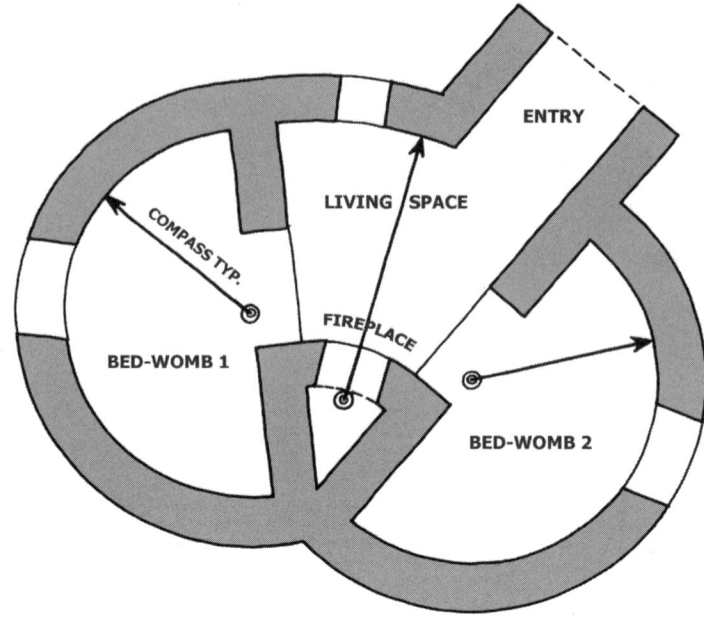

This structure is a playful adventure with arches.

MODEL 9: Homeless Deluxe

The next dome, the Homeless Deluxe is a two person shelter of less than 80 sq. feet. It is a more complex shape than the rest of the village, but because it is small it is not hard to build. The apses/pouches are attached onto straight walls, facing each other and leaving a triangular common area between them. These are two private sleeping rooms (bed-wombs) which are entered from a shared living room with a triangular fireplace, which heats all three areas.

The two rows of bags and barbed wire below the floor level are continuous, following the principle of two complete rings for a Super-adobe foundation. A 14 or 16 inch wide bag may be used.

Right: Foundation rows and dome layout.

Left: Building the lower Superadobe rows.

Fireplace

The walls of the foundation and base rows are integrated with the triangular corner fireplace. A form is set up for an arch over the fireplace hearth, and a pipe is built in at the back to bring in fresh air for combustion. To truly understand fireplace design you should study traditional fireplaces, which can be copied and often simplified for a temporary emergency shelter.

Since earth construction is non-combustible, the heat from the fireplace can transfer into the walls and warm the building. (In buildings of combustible materials, fireplaces need insulating). The firebox should be lined with earth/clay plaster.

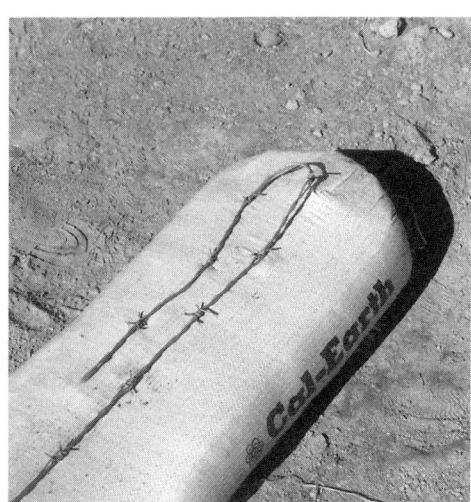

Above: Where the wall ends for a doorway, the strand of barbed wire is looped back for 2 ft. to provide extra grip in these weaker areas.

Left: Setting the fireplace arch form on wedges and inserting a vent pipe.

SUPERADOBE - SANDBAG SHELTER

Above: Compass chain in use.

Below: Filling the bag end on the metal plate or "pizza pan".

This little Homeless Deluxe shelter has three compasses whose locations are in three areas, for building the two apses and the triangular room, which is a partial dome. Each apse has a compass in the center and a height compass on the circumference of its projected circle. A third center compass chain is attached on top of the fireplace arch to make the partial circle for the triangular living room. Its length is determined relative to the apse walls. When the upper rows of this shelter are reached, only the living room compass is needed. In this dome all the compasses were made using the two-chain method (see Appendix IV, Compass).

Bag Ends and "Pizza Pan"

The filled bags end at the various entryways, and are neatly folded under and tamped. To accurately position the bag endings at the entry-ways, without fighting with the barbed wire, a helpful tool is a smooth plate of metal or wood (a "pizza pan"). The entryway bag is filled on top of the smooth plate, which insulates it temporarily from the barbed wire. The builder freely adjusts the bag position during construction, then the smooth plate is slid out to allow the barbed wire to grip. Finally, the bag is tamped permanently into place. The end of any row may be compacted using a brick, or packed by hand.

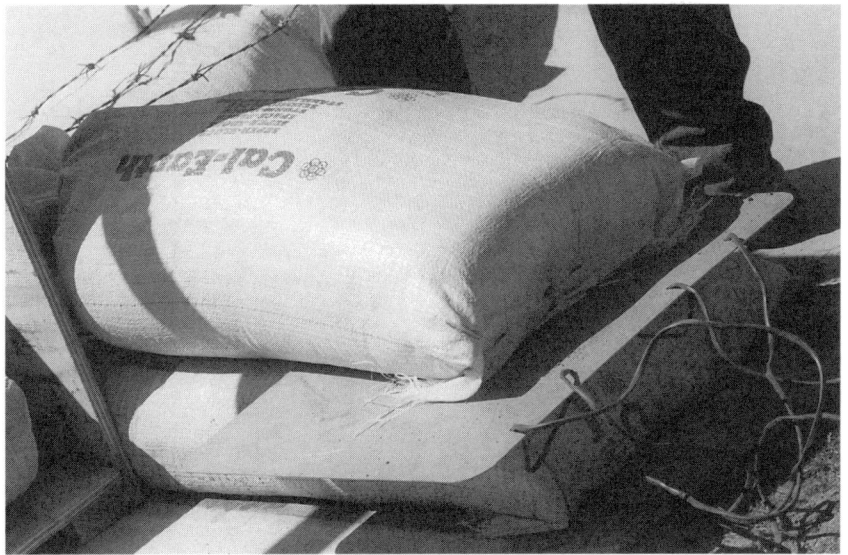

Left: Using a metal plate to position the bag ends.

Above: Barbed wire loops for attachments.

A carpenter's level or a plumb bob is used to check the vertical walls, especially at the entryways.

The barbed wires can be extended out beyond the wall where attachments are planned. Short pieces, such as the loops shown on the right, can attach finishing materials such as thick plaster, eyebrow elements or brick and stone veneer. Longer pieces of two to three feet or more can attach another wall or dome cluster.

Below: The fireplace hearth arch, and compass chain attached to it. Recycled concrete blocks are used to locate the temporary arch forms.

SUPERADOBE - SANDBAG SHELTER

Above: Packing the end of a row by hand.

Right and below: Overlapped barbed wire connections and interlocking bags. The dome has many walls which join.

More on Connecting Walls

The apses/niches and spaces of the Homeless Deluxe are closely connected and eventually merge at the top into a single domed form.

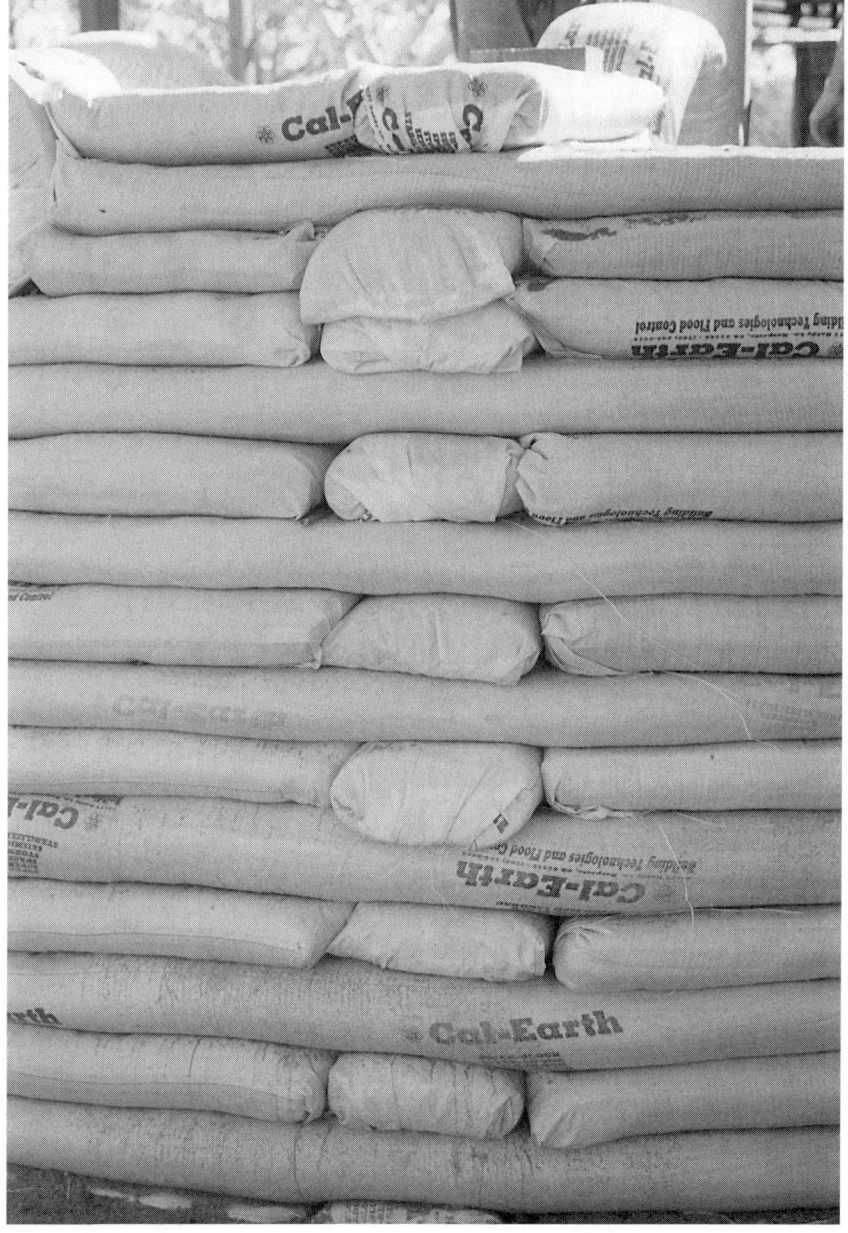

Above: Gloves should be used to handle the barbed wire. A fence plier is useful to push in stray barbs.

Left: Detail of interlocking bags where the walls join. (Interlocking in all these small structures is not necessary since we are using continuous barbed wire connecting them together).

    The apse/niche and the dome wall are interlocked by overlapping alternate rows, like traditional brickwork (every three or four rows is sufficent). The four point barbed wires must be continuous or, if the wire is cut, the two ends must be overlapped by at least two feet and twisted together, as is standard throughout all these buildings.

    Interlocking bags are used here because the apse and dome elements merge together into a unitary structure. However, a cluster of repeated elements which do not merge, such as the Caterpillar/Sinapsoapsis segments, or a series of domes connected by small entry vaults, should be connected by the barbed wire alone. This allows each segment to move separately, if needed during an earthquake.

SUPERADOBE - SANDBAG SHELTER

Right: The wooden window formwork is inserted.

"Eyebrows" With Long Bags

The following techniques, using long bags to create window eyebrows which protect from rain, are recommended for stabilized earth only.

Using a wider bag creates the window arch and eyebrow together, in cases where a wider bag is available. Here, a 26 inch wide bag was pinned with nails to tailor it into a lozenge shape and draped over the arched formwork with the wide part at the top. A slit was cut at the top of the bag, through which the stabilized earth was packed by hand down both sides of the arched formwork to make the arched shape.

Left: A wide, tailored bag makes both the window and eyebrow. Note the earth-filled hole in the top of the bag.

Below: The form is removed from the wide, tailored arch. Two strands of barbed wire join the arch and the dome bags.

After filling, the arched bag must be compacted well with a brick or tamper. This wide, tailored bag has a base of 16 inches wide to fit the 16 inch dome wall, and a top which flares out to about 26 inches wide creating the "eyebrow". The inner edge of this bag arch curves with the inner surface of the dome wall.

The window arches are buttressed by rows of bags which make the dome wall. The size and shape of the arch and the type of stabilized earth determine how many rows of bags must be completed before the arched formwork can be removed. By using cement stabilized earth, the form above could be removed after four buttress rows.

SUPERADOBE - SANDBAG SHELTER

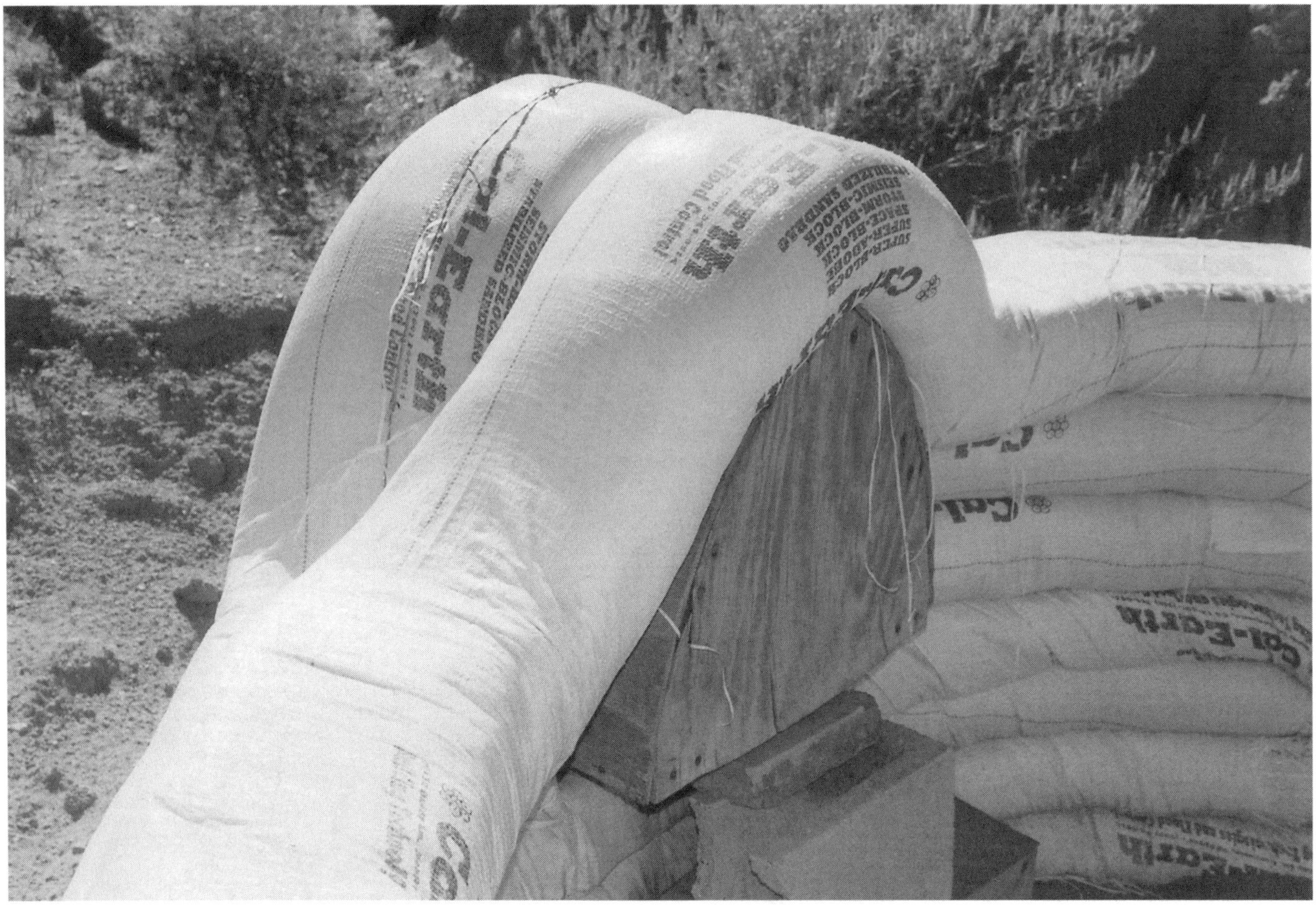

Above: Eyebrows can also be made from long bags laid side by side.

Below and right: Long bag eyebrow using a bag form, instead of a wooden form.

The most commonly used eyebrow technique does not need a wider bag, or tailoring. Two rows of bags are laid side by side over an arched form. The outer bag projects out and forms the eyebrow, whilst the inner bag forms the arched window in the dome wall.

Left: Packing and tamping the long bag with a brick.

Above: The builder supports the long bag on his leg and body until it is full enough to complete the arch.

Below: The inner bag goes over the form.

1) This is a minimalistic, creative and organic technique, following the principle of the way some sea creatures make their seashells. Just as the sea creature utilizes a continuous curving flow of material to make its shell, so the "seashell eyebrow" can be built without interrupting the work flow.

2) The inner bag forming the window arch is the first to be laid over the arched formwork. It must follow the inner curvature of the dome wall, and its location is checked using the compass.

To make an arch with a long bag, the first half is filled resting on the formwork and shaped with a brick at the same time. For the second half, the builder uses their leg and body to support the bag while it is being filled. When enough earth has been packed into the long bag, the builder twists the bag material to temporarily close and, without letting go, lowers the filled bag over the arched formwork, while staying balanced and controlling the position of the twisted end to set it onto the dome wall and the barbed wire.

Immediately, the arch is firmly shaped and compacted with a brick or tamper to become a self-supporting structure. The builder continues to fill the rest of the long bag so that the dome wall and window arch are continuous.

Right: Barbed wires are attached to the inner bag and left sticking out.

Right: The outer bag for the eyebrow is filled and laid over the form.

3) Barbed wire must tie together the inner and outer bags. In addition to the continuous barbed wire which runs parallel to the long bags around the dome, several strands of barbed wire are wrapped around the inner arch and left hanging out from the dome at right angles, as "tie-back" strands. Later, these tie back strands of wire will wrap around the outer bag tying it back onto the inner bag and weaving into the dome wall.

4) The outer bag will become the eyebrow, and starts by being part of the dome wall. The builder continues filling the long bag resting it on his leg and torso as before, and using the inner arch to support it. After twisting the bag closed and lowering it down resting on the inner arch, the outer bag is gently pushed or massaged sideways so that it slides down onto the formwork and fits tightly next to the inner arch. There should not be any gap beween these arches, although they may overlap somewhat.

Left and below: The outer arch is tied to the inner arch using the barbed wires.

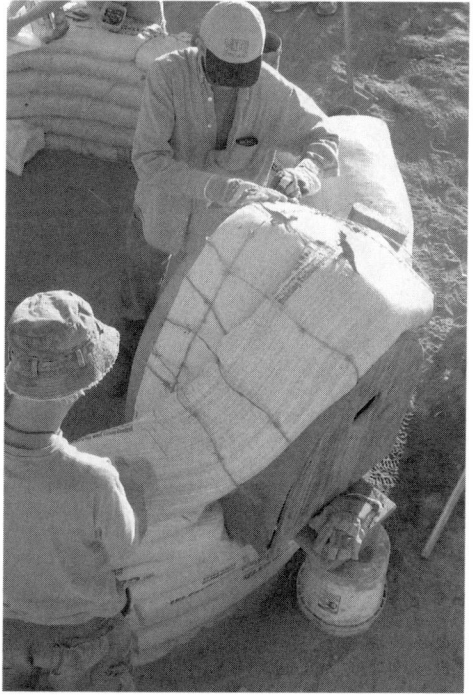

SUPERADOBE - SANDBAG SHELTER

Right: The continuous Superadobe arch is being built with the help of a wooden board for temporary support. The bag is being filled using a tube.

Below: The board is removed and the bag is laid over the arch onto the wall beyond.

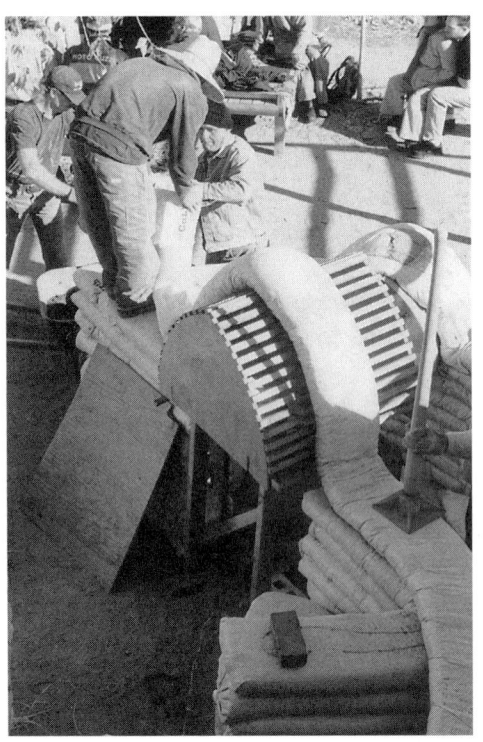

In the Homeless Deluxe, the openings into the apses are built as arches over wooden formwork, without needing eyebrows. Going up the form is easy since gravity is on your side, but coming down needs support. For this larger arch, a piece of wood is used instead of the builder's body, to support the bag. When enough earth has filled the bag as a long column, the builder twists the bag material to close the end tightly. The wood support is removed and the filled bag is carefully lowered over the arch form. If more length is needed, the twisted end can be gradually opened to allow more earth to slip down and, when the builder is satisfied that the arch is correctly placed, the whole Superadobe arch is tamped into place.

Left: The arch formwork here is supported on wooden legs, and uses small wood strips making the form lighter.

Entry Vault of Arches

The main entry to the structure is a small vault, made by repeating Superadobe arches, which are connected into a vault by weaving barbed wire above and below. For a larger vault, steel bars or mesh should be used instead of barbed wire.

Below: The bags for the entry arches are twisted closed, lowered over the arched form, and shaped with a brick.

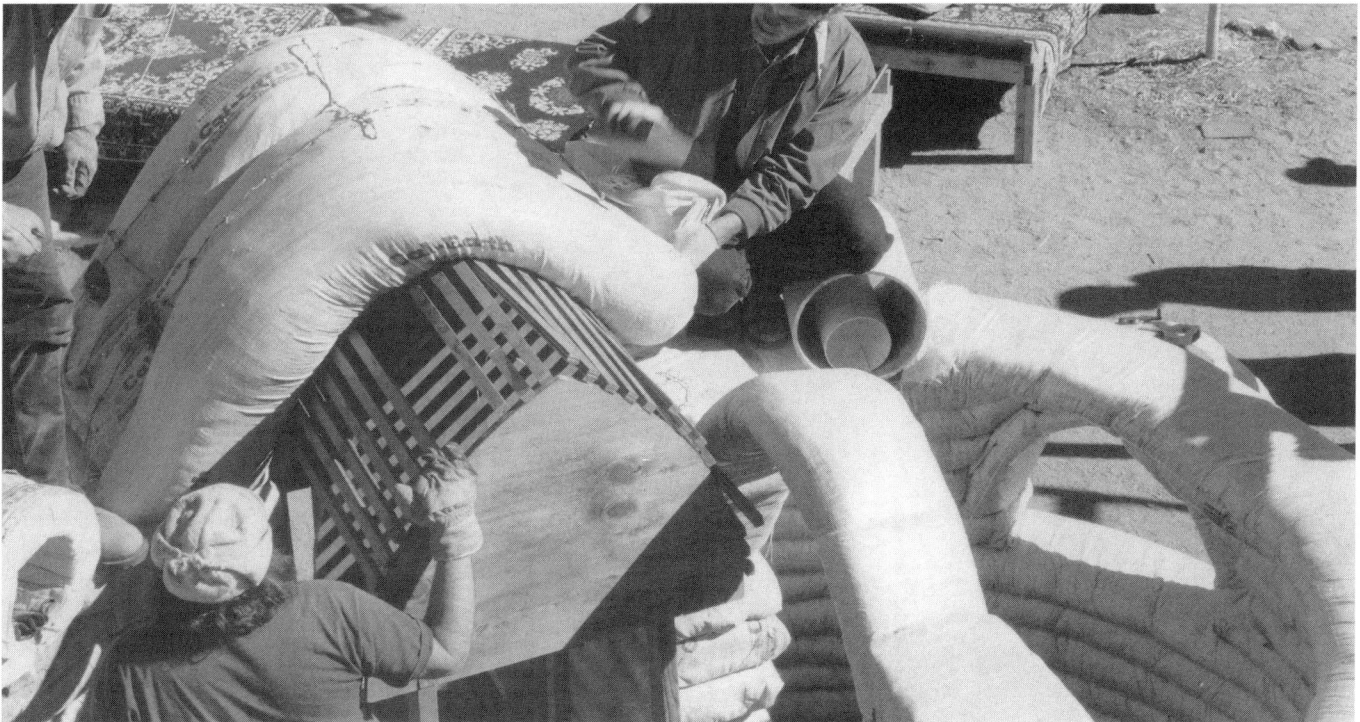

Right : Multiple arches of stabilized earth used for the entry vault, the doorway into a niche, and a window.

Below: The arches of the entry vault, the doorway into a niche, and the fireplace.

In this example of a stabilized earth dome, we can now see that the long bag arch has been used in several locations with some small differences that are appropriate to its function.

The window arch was built using a wider bag, tailored to create an eyebrow. Two long bags laid side by side can create the eyebrow and the windows differently. A single long bag arch makes a doorway in an interior vertical wall, where no eyebrow is needed. Several long bags make a small vault for an entryway, and these must be tied together horizontally with barbed wire. A long bag arch also creates the fireplace hearth. These are five different examples of arches.

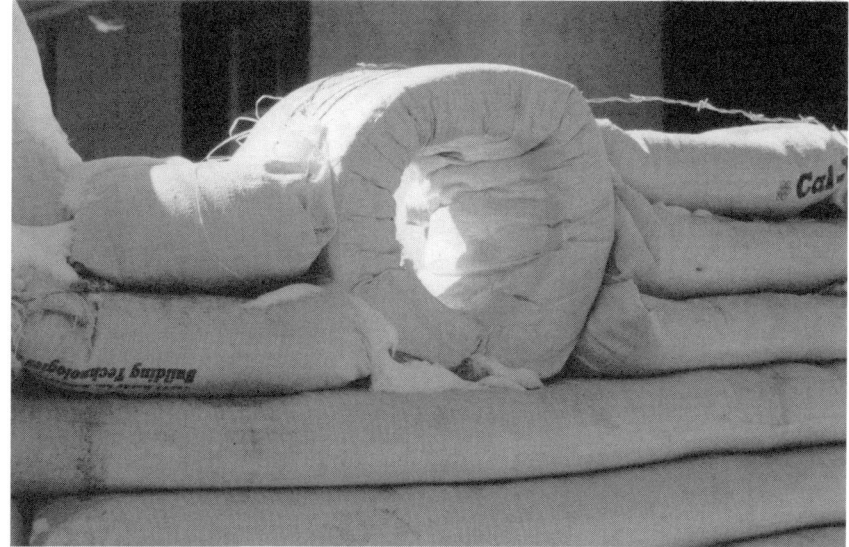

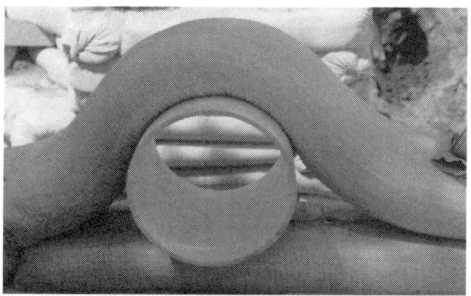

Above: A pipe forms a window.

Left: A pre-cast round window made from a stabilized earth-filled bag.

Circular windows can be included in the design by pre-casting on the ground before lifting into place. Regular pipe windows of small sizes are also used.

Below: Multiple arches and pipe windows.

SUPERADOBE - SANDBAG SHELTER

Above and right: The rows of bags continue up to the top of the dome.

Free-Style Corbelling

After several rows of bags have been completed above the window and door arches, the builder is now able to walk freely over them. The apses and walls become convenient places on which to stand while building further up.

At the top of the Homeless Deluxe, the two apses and the central living room were gradually merged by corbelling inwards a little at a time. This unified the top of the structure.

Because of the unusal shape which closes the structure, the upper Superadobe rows were built carefully and slowly enough to allow

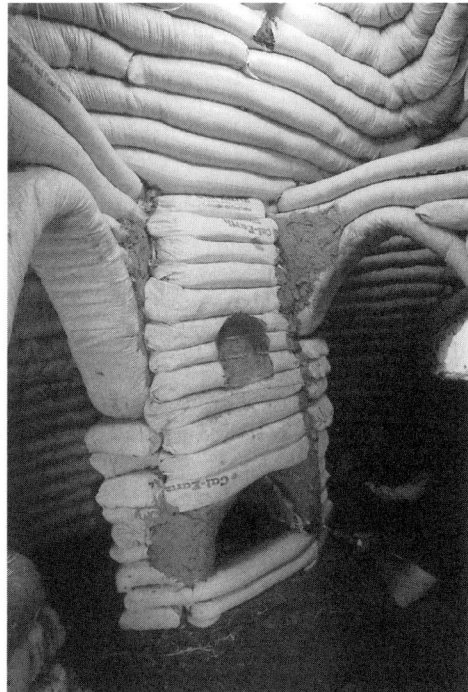

Above: Dome interior showing the fireplace, niche entrances, and coils above.

Left and below: The top of the structure is gradually corbelled inwards to become an ellipse.

them to dry and set hard, before building more rows. The builder was experienced enough to use his judgement to step in the coils without needing a compass.

The following four points made this free-style merging a success:
1) The dome is small
2) The earth is stabilized.
3) The approximate shape is an ellipse.
4) The eventual opening at the top is round.

The small round opening at the top will become a skylight and could also function to vent the rising hot air in summer. The top of this dome could also be closed with bags.

SUPERADOBE - SANDBAG SHELTER

Right and below: The coils of bags gradually corbel inwards above the doors and windows.

Looking from inside, you can see that the bags are stepped in, or corbelled, fairly evenly above the doors and windows towards the top of the dome. The builder may also shape the inside surface of the dome with a brick during construction to create a smoother curve.

Interior Plaster

The interior gets a coat of plaster, here applied by hand. The coil shape of the bags is even more important inside than outside, in helping the plaster to adhere and "key" into the walls. Wetting the wall before plastering is necessary, and removing the bags also helps the plaster

Right: Applying mud-straw plaster onto the interior of a dome.

Left: Interior view of plastered walls and fireplace on the left.

Above: Interior plastering with stabilized earth plaster. Traditional plastering has been done for thousand1s of years by women's groups who could socialize while working.

to adhere better onto the upper surfaces.

In general, the interior plaster does not need to be water resistant if the exterior finish is completely waterproof.

A traditional mud-straw plaster sticks well to the underside of the Superadobe rows. This contains clay earth, sand, and straw, mixed thoroughly with water and left to sit for at least three days to release lactic acid, a binder and strengthener of the plaster. A mud-straw plaster of one third clay to two thirds sand and silt, with the same volume of straw added is a typical proportion. Before using, the plaster must be kneaded, traditionally by foot, to bring out the plasticity and strength of the clay and create a cohesive plaster.

If dampness affects the interior, a stabilized earth plaster using lime or cement is necessary. A cement stabilized plaster will only stick to the upper curves of dome if a wet slurry is applied in thin layers. Lime stabilization makes the plaster adhere better.

Right: The vertical walls around the arch are plastered with smooth stabilized earth, and the curved interior of the sleeping niches is painted with white stucco, after the bags are removed. The stucco is also used to highlight the archway. The wooden panel in the floor is to cover the storage underneath.

Here, it was decided to use stabilized earth for a smooth hand-finished plaster on the vertical and lower wall sections .

And for the upper rows, if there is no need for a smooth interior plaster, the bag coils can make a beautiful finish. Here, white stucco has been mixed to a slurry and smoothed by hand, or painted on with a brush, adding both color and a long lasting finish.

Storage and Experimental Heating

The sleeping areas are raised above the floor level. The bedding is placed on top of a wooden access panel to a storage space underneath.

From Iran to Japan traditional people have kept warm sitting at a low table ("korsee") with a heating element underneath, and a large quilt draped over themselves to the ground. Transferring this concept here, one or two incandescent light bulbs in the storage space under-

Left: Fitting the access panel to the storage area under the sleeping niche. This space is also utilized to warm up the bedding over it.

neath, radiate their wasted heat into the walls over time, which sends warmth through the wooden panel under the bed area. Additionally, the fireplace brings warmth from its walls into this chamber. The same principle could apply for a windscoop to bring in cool air.

Left: Sleeping is comfortable in this "bed-womb" warmed from below.

Above and right: Completing the rough plaster coat over the bags.

Below: The rough plaster is complete and the dome is ready for the waterproofing.

Waterproofing Layer

As a general rule, waterproofing layers must follow the available local waterproofing methods that have proven themselves over time.

To be effective, a waterproofing layer must:
a) be applied to the surfaces where rainfall can penetrate
b) be flexible for expansion and contraction with the building
c) adhere well to what is under and over it
d) be long lasting
e) be easy to repair or patch.

For this dome, asphalt was chosen (also called tar or bitumen) as a traditional, available waterproofing material. The fabric of the bags also became part of the waterproofing system.

1) A scratch coat of exterior plaster was applied first, filling the coil shape which gives a good "plaster key". This can be spread directly over the bags since the curves of the dome help the plaster to adhere to the wall, making its own structure.
2) Two coats of liquid asphalt are then trowelled or brushed (cold application) or hot-mopped over this, generally on the upper surfaces and gutter areas.
3) The first primer coat is usually more fluid and soaks into the plaster, bonding well with it.
4) A second, thicker coat, is spread over the primer coat.
5) Next, fabric strips are pressed onto the asphalt to make a reinforced layer. The strips overlap to cover the entire roof area. The sandbag material itself, when cut open, can be used as the fabric. This will resist the downward creep of the asphalt over time, and prevent cracking from settlement.
6) The fabric layer is then covered with a third, thick coat of asphalt,

which is sprinkled with sand to create a surface for the finish plaster.
7) Finally, the asphalt must be protected from sun with a layer of finish plaster, such the Reptile finish or smooth stabilized earth.

The finish plaster chosen for this dome was smoothly trowelled stabilized earth, in a thick coat over the fabric. The dome was plastered in smooth contours to allow all rainwater to flow away from the building, and for snow to shed easily off the steep slopes.

In a wet, cold climates water must not collect on the surface to avoid freeze/thaw damage. Porous, breathable surfaces, such as lime-sand plaster, are traditionally used to resist frost better than denser ones, since denser materials will hold water for longer periods of time.

Some liquid waterproofing alternatives to asphalt are as follows:
1) A thick clay-straw plaster may be used in a dry climate as both the scratch coat and the waterproofing layer, provided that an erosion and crack-resistant finish plaster is added on top, such as the Reptile mud-balls, or a lime-based plaster.
2) If manufactured sealants are available, these can simplify the waterproofing in many climate zones since they can be painted, mopped, or otherwise directly applied onto the finish plaster, following the manufacturer's instructions.

Above: Painting liquid asphalt layers onto the stabilized earth rough plaster.

Below: Pressing strips of fabric into the asphalt.

SUPERADOBE - SANDBAG SHELTER

Above: Adding a layer of asphalt on top of over-lapping vertical strips of bags.

Right: Finish plastering over the waterproof membrane, with a steel trowel.

Below: The last layer of thick liquid asphalt has been applied over the fabric. It is dusted with sand for the finish plaster to adhere.

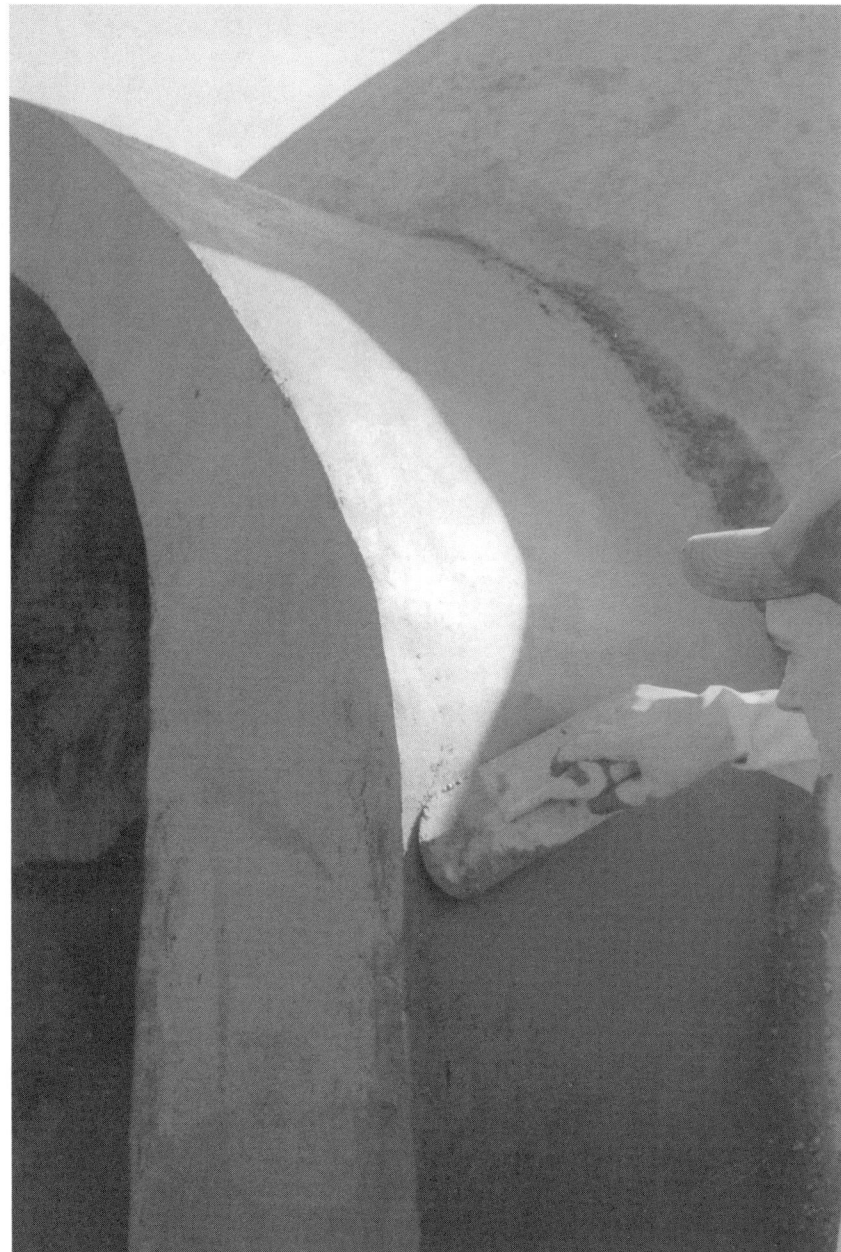

Finish Plaster

A steel trowel can create a durable exterior finish. Using the trowel to work and press the plaster brings the cement to the surface and drives the larger grains of sand and gravel down beneath the surface, resulting in a harder, erosion resistant finish.

The trowel used here is a steel "float" commonly used for stucco work or concrete slab finishes. Flexible steel with rounded ends is best for plastering the rounded form of the dome. (Trowels can also be

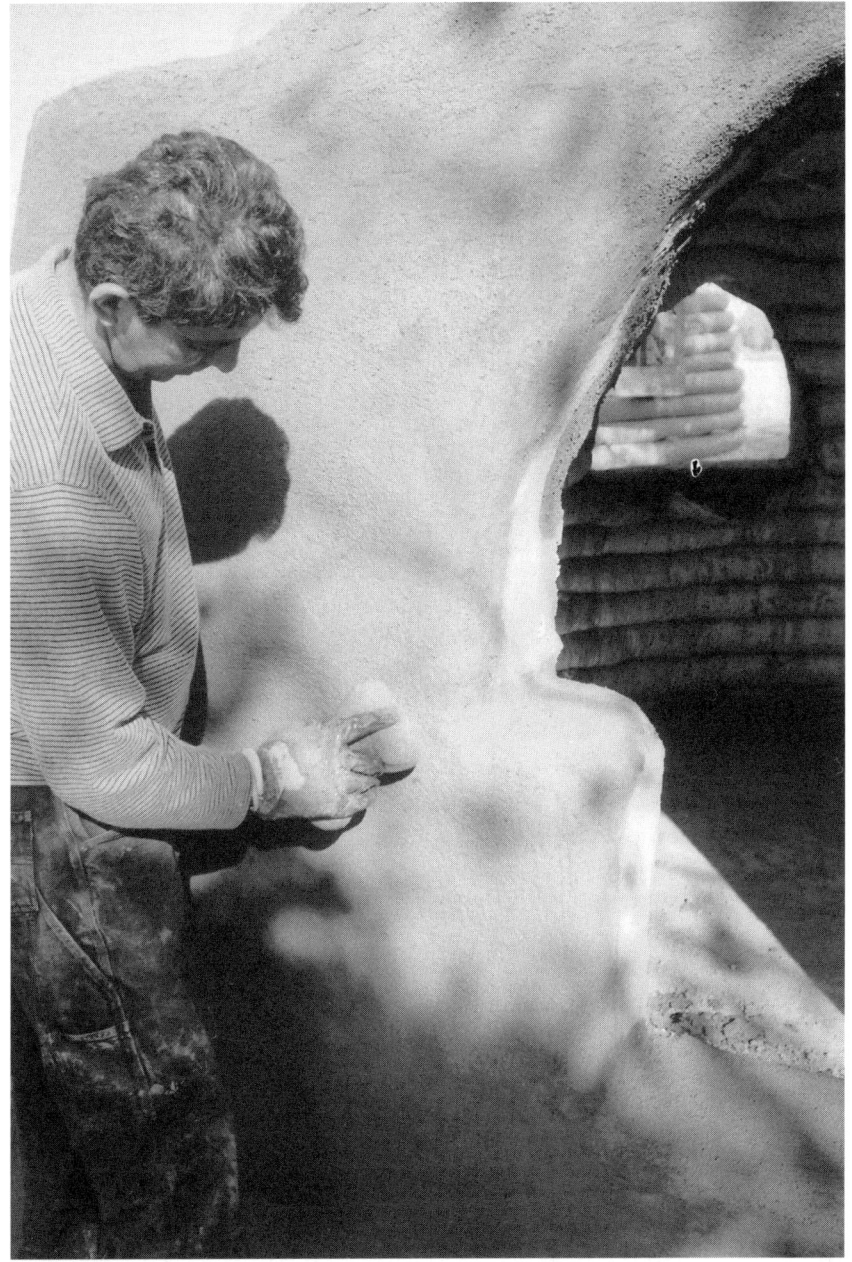

Above: The finish plaster layer is contrasted with the rough coat for aesthetic effect.

Left: Finishing with a damp sponge.

Below: Smooth sponge finish around an inset pipe window creates soft and more desirable features.

made using recycled steel or plastic in rounded forms).

A steel trowel gives the option of finishing with sharp lines and edges, as well as smooth rounded corners. Since a finish plaster with a soft texture and rounded edges is more desirable, the builder uses a damp sponge to lightly rub the plaster before it hardens (at the leather hard stage within an hour). The sponge also smooths any hairline cracks which may develop as the plaster cures and dries.

Windows must be installed before plastering, and must be sealed using a waterproofing method such as asphalt. The plaster then surrounds the window. Small or large eyebrow elements sculpted by

Right: Sponging the finish coat on top of the rough plaster.

the finish plaster are also part of the waterproofing.

If doors, windows, and a skylight are to be installed, these are also done before the finish plaster stage. The overall form of the plaster should create a smooth, steep slope for rain water to flow quickly

Below: The finished dome uses cut lines and changes in texture aesthetically.

Above and left: Sculptural finished form and texture around a window.

Below: A steel and glass openable window fitted into the arch, sealed, and plastered.

down and away from the dome, and for snow to shed easily.

Often it is not possible to plaster the whole dome together. Therefore, a cut line is left around each finished section. Later the builder may blend together the new and the old plaster, or he may leave the line. Here, the lines were left as expansion joints and were artistically included into the plaster pattern, along with variations in smooth and rough textures.

Thus an emergency shelter of Superadobe sandbags and barbed wire, can be upgraded with plaster, waterproofing, flooring, a fireplace or windscoop, doors and windows to become a long-term structure. Whilst the upgrading shown for the Homeless Deluxe demonstrates what is typical for an emergency shelter, it forms the basis for the techniques used for a permanent home such as the Eco-Dome. Areas of further upgrading include plumbing, electrical, more extensive waterproofing for foundations, additional roofing, flooring and interior fittings.

Right: A strudtural adventure through spaces, arches, windows, and finishes, for people of all ages.

Below: Finished Homeless Deluxe in the snow.

MODEL 10
SEASHELL DOME

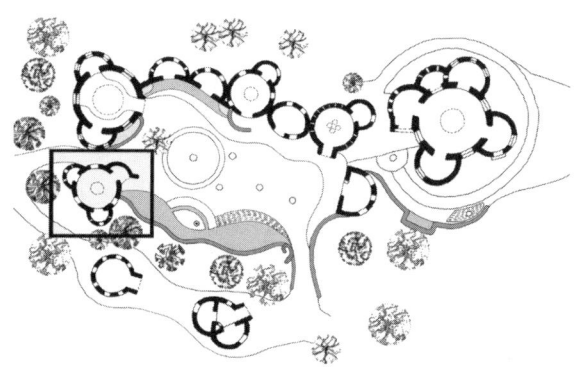

SUPERADOBE - SANDBAG SHELTER

Right: Plan of the Seashell Dome. The dome and each apse/niche have a compass.

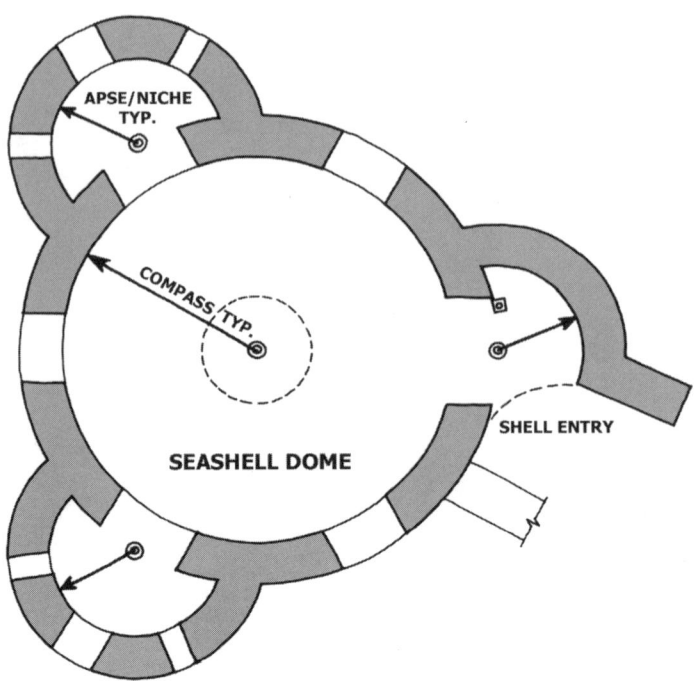

MODEL 10: Seashell Dome

The seashell dome gets its name from its coiled form, the curved entrance apse and its round coiled windows. It learns a design lesson from seashells. The sea creatures who only follow their genetic instincts make the most beautiful homes with the best forms, colors, texture, and even waterproofing, all from the seawater itself.

Right: Forms of seashells can be design inspirations.

Left: The walls and curved entrance apse are started below grade with a stabilized earth foundation and stem wall, followed by two Superadobe coils, also below grade.

Below: The temporary arch forms are set into the dome. Using strips of wood in their construction makes the forms lighter.

SUPERADOBE - SANDBAG SHELTER

Above: The dome has several openings demonstrating the flexibility of the Superadobe construction technique, which makes sculptural forms easy. All the lines are organically curved.

Below: To build the circular windows, the Superadobe bag was filled with stabilized earth and draped around a form. The builder uses the same technique as for the entrance arch and windows.

Above: The builder stands on top of the walls while he builds. The wooden forms are deep enough to allow an eyebrow projection to be added over the window.

Below: More bags for the walls are immediately laid between the arched openings to act as buttressing, and a minimum of three feet of solid wall is built between the openings.

SUPERADOBE - SANDBAG SHELTER

Above and left: After at least four rows are completed above the windows, the temporary formwork is removed.

Below: View into the dome and niche showing the continuous Superadobe rows which build the wall, arch, and the round windows.

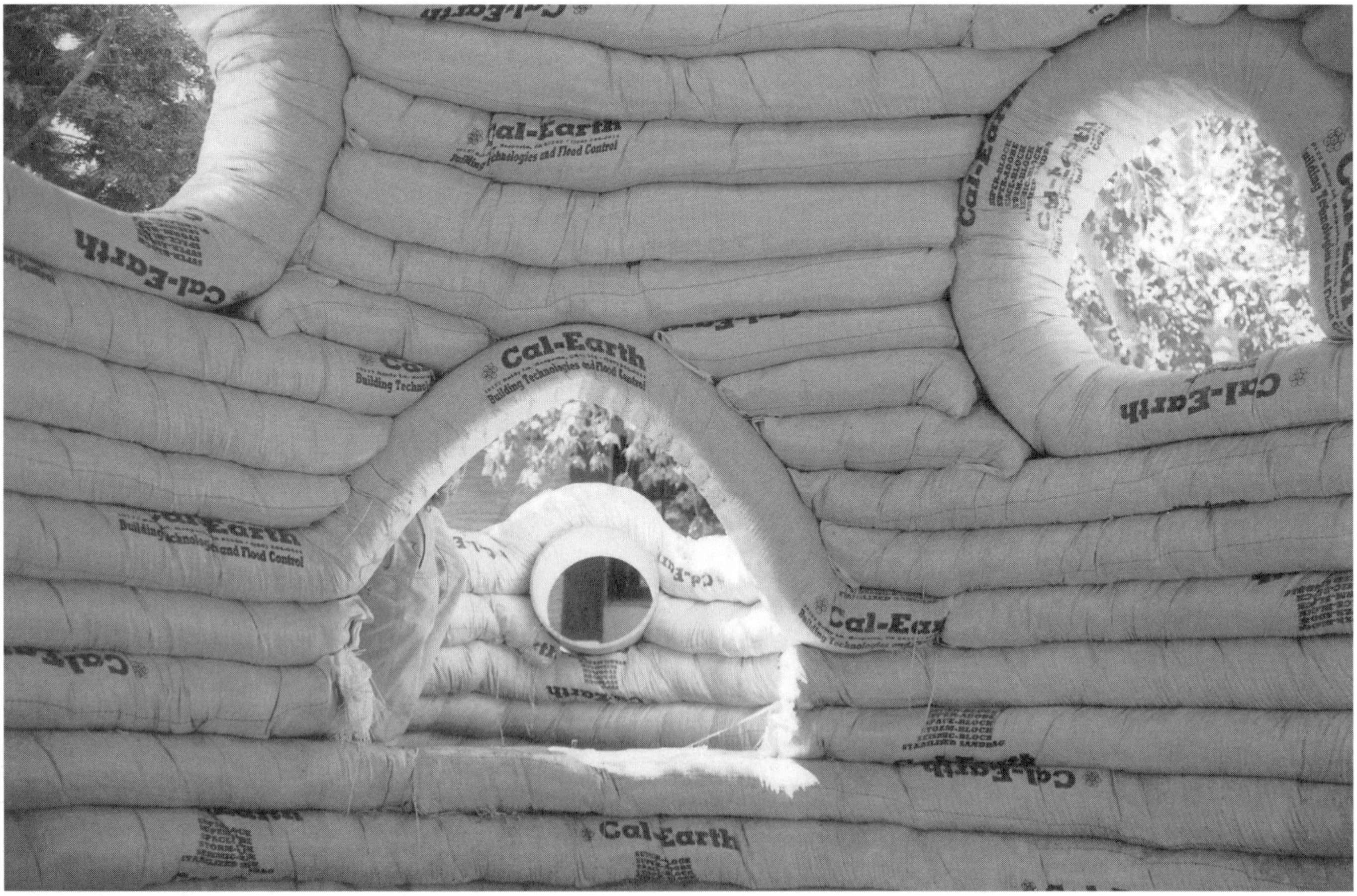

Above: A small niche is constructed after the main dome, connected to it with barbed wires. A 12 inch diameter pipe window is inset into the niche wall.

Below: The niche is just large enough for a child's cot or sleeping place, or a utility space. To add the niche later leave a knock out panel in the dome wall.

SUPERADOBE - SANDBAG SHELTER

Above: Tamping the upper coils.

Right: Placing the barbed wire to connect the dome and entry apse.

Below: Finishing the upper dome with the entry apse.

Shell Entry

The dome and entry apse were built at the same time and were connected with barbed wire and interlocking bags. Since the upper part of the entry wall is straight and heavier at the top, its stability is created by the following four factors:

1) The entry apse is small and the walls are relatively thick.
2) It is built with stabilized earth.
3) The Superadobe rows on the entry apse are connected into the main dome with barbed wire and interlocking bags.
4) There is only a short piece of free standing straight wall.

The top of the Seashell Dome is closed with coils like the other domes in the village. The entry apse, however, is closed at the top using the leaning arches technique. Leaning arches for an apse are similar to those for a vault. The first bag is short, then a longer one is pitched over it ond so on. The purpose is to create an arch span which is partly supported by the bag below during construction. Each arch leans on the one before to complete the entry.

On a small dome like this one, an experienced builder can gradually corbel the rows inwards without a compass.

Above: Detail of the entry wall and leaning arches, completing the apse.

Below: The curves of the shell entry apse. Leaning arches create the apse form.

SUPERADOBE - SANDBAG SHELTER

Right and below: The completed entrance, inviting and protecting.

    The entrance is a welcoming curve leading us out of the harsh winds, and into the beautiful light and airy interior of the Seashell dome.

    If a little sea creature, who does not claim as we do to be created in the image of God, can make a home which is the best shell structure, best patterns, colors and texture, a home which is in harmony with the ocean environment and which is all created from sea water and waterproofed with water, then why shouldn't we be able to pick up the earth and build a home for ourselves that can resist the elements and works in harmony with nature?

Above: Burning off the top and bottom of the bag in a line, is an option for exposing the coils.

Removing the Bags

A torch was used to remove the bag material. The wall was then soaked with water before applying stabilized earth plaster, which was similar to the wall material for equal expansion/contraction.

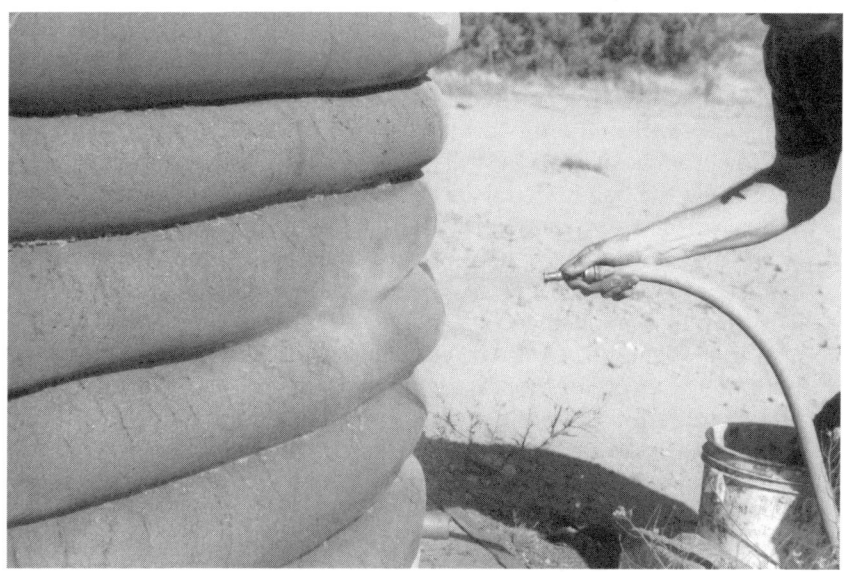

Left: Soaking the wall with water before plastering.

Above: Starting a minimum finishing on the exterior of the Seashell Dome.

Minimum Plastering

To express the coil lines without covering them in thick plaster can be accomplished if the following conditions are met.
a) In dry climates where rainfall occurs between long dry spells, the thick earth walls act like a slow sponge, soaking in the rain and drying out when the sun shines. Under these conditions a thin exterior finish with minimum waterproofing is sufficient. Generally, the lower areas of vertical walls do not need waterproofing.
b) Removing the bag material and soaking the wall with plenty of water is necessary when applying the exterior finish; here a stabilized earth plaster is applied directly over the structural coils of stabilized earth, so that the two materials are similar in nature.
c) This thin layer may also contain a waterproofing compound; for example a waterproof stucco containing manufactured products or natural additives such as resins, acrylic, enzymatic stabilizers, oils and waxes added into the plaster. Traditional recipes include cactus juice, and a highly polished finish using lime, fine silt/clay and sometimes ash.
d) To delay water absorption, water-resistant fluids such as clear sealants, soil stabilizers, waterproof paints, asphalt emulsion, or latex-

Above: Finished Seashell Dome.

Left: Applying a thin layer of exterior plaster and giving it a smooth sponge finish.

based sealants which penetrate the stabilized earth surface can be painted over the coils. Additional layers such as white lime or stucco can also help to slow down the water penetration. These can perform as a primary or secondary water resistant layer on any parts of the dome which receive extra rainfall; for example the top of the dome or any flat areas, the window sills and eyebrows, and the connection of the dome and apse where a natural gutter is formed.

Thus a water resistant or waterproof finish can be directly repaired or renewed as needed, provided it is applied as the outermost layer. The entire exterior can also be decorated with indigenous painting and color finishes.

Right: The Seashell entrance celebrates the coils giving an impression of depth and perspective, and invites one to enter the security of its interior.

Below: The infinite expressed in a curved vista.

The final appearance of the Seashell dome celebrates the lines of the undulating shell-like coiled exterior. It has only a thin plaster, waterseal and accents of white lime stucco.

By designing with the curve of the buildings and the coils themselves, the sense of the infinite, of long vistas and open spaces can be created through the art of architecture. The open horizons given to us by nature are usually lost, and missed, when buildings rise up from the land. Inspiring architecture can give back to us this sense of freedom that comes with the natural landscape, while keeping us hugged within the close curves of the urban landscape. Thus the layering of curved surfaces over a short distance invites the impression of an unconstrained landscape within the walls themselves.

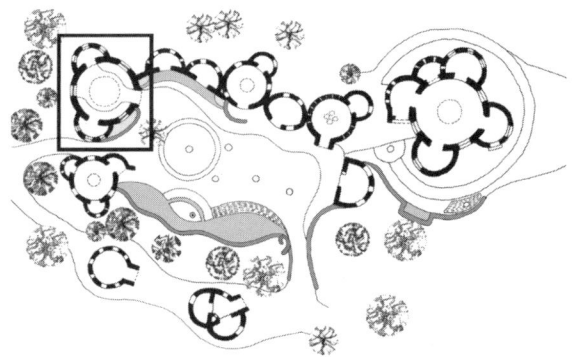

MODEL 1
ROOFLESS DOME

SUPERADOBE - SANDBAG SHELTER

Right: Plan of the Roofless Dome cluster, with a cylindrical central structure.

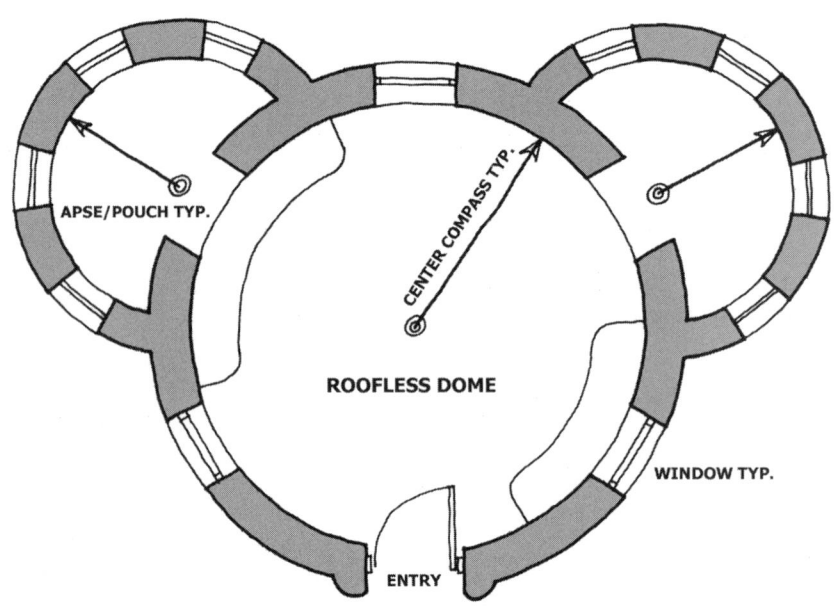

Below: Roofless Dome under construction.

Left: Interior of the Roofless Dome.

Below: Finished entry and exterior reptiles.

MODEL 1: Roofless Dome

The last dome in the village of prototypes was also the first to be built. It is the simplest because it is roofless. It is a cylinder of Superadobe bags that does not close into a dome. Instead, the open top is covered by a shade surface of reed mats and recycled netting. Inside becomes a shaded private courtyard enclosure, open to the sky, but protected by the thick Superadobe walls. On its sides are two pouches (or bedwombs) which are 8 ft. diameter apses. These are coiled until they close at the top, then waterproofed and finished like the other pouches of the village.

SUPERADOBE - SANDBAG SHELTER

Right: The windows and door frame are built directly into the Superadobe walls, including knock out panels for the apse openings.

The construction consists of stabilized earth Superadobe bags for the foundations and walls. A dry well was dug out in the middle of the floor to provide drainage during rain showers. The 8 ft. apses were attached to the main dome with barbed wire.

The springline for the dome center compass was set at 3 ft. high

Below: The final coils are placed on the dome.

Above and left: Details of construction at the top of the dome. Special care was taken to stabilize and tamp the final coil which will be wider to create a drip edge.

resulting in the more cylindrical shape, and because of this shape the entry door does not need additional buttressing. In general the height compass was not needed as much in this dome, since the builder's judgement combined with a center compass were sufficient for the roofless dome and small apses.

Window and door construction includes:
a) knock-out panels of filled bags for the apse entrances
b) round pipe windows
c) wooden arch forms above the door and square windows
d) windows built directly into the wall without frames, and positioned for ample light and cross ventilation

Below: A knock out panel of dry sandbags being removed for the apse opening.

Left: Construction of the apse over arched and pipe windows.

SUPERADOBE - SANDBAG SHELTER

Above: Adding a temporary "shower cap" and tying it to the bars/pipes.

Below: Top of the apse and dome.

The reed roofing mats for shade are attached to anchor points on the dome perimeter by ropes, and recycled netting is stretched underneath for support.

When additional protective covering is desired because of rain, a temporary "shower cap" tarp may be stretched across the open top for the winter or rainy season. It is tied down to a dozen short steel rebars and pipes on the perimeter of the dome, which are sandwiched between the upper Superadobe coils. A frame of flexible bamboo or steel/plastic pipes can be used to give a pitch to the tarp making it more like an umbrella. In the dry season the tarp is removed and stored in an apse.

Even a permanent roof may be attached to the top of the dome, like a hat, provided that it is lightweight. The concept is similar to the added roofing for tropical climates, using palm leaf "Palapas", thatch, corrugated sheeting, or other available materials.

To support a heavier roof the top coil of the dome must be reinforced as a bond beam and the top four coils more must be strapped together. Heavy roofing is not suitable in high seismic areas.

Since the goal of the roofless dome is to provide a design for dry climates that is simpler to construct than a complete dome, the additional covering elements should be kept simple and lightweight using appropriate local materials.

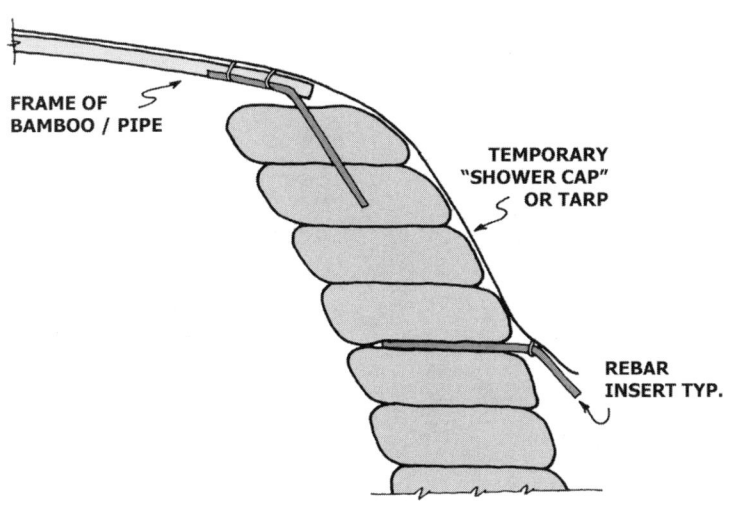

ROOFLESS DOME - DETAIL AT TOP

Below: Reed roof mats cover the dome top. The frame above has no structural function. The short rebar are anchor points for a temporary "shower cap".

Above: Reptile plaster is applied over the apse, onto asphalt waterproofing.

SUPERADOBE - SANDBAG SHELTER

Above: A knock out panel of temporary dry sandbags will make the apse entry. Its shape is later corrected with thick interior plaster.

Right: The light and shade of the roofing create a cooling effect in the enclosed spaces. The central space can be used as an open living room.

Below: Plastering the interior by hand.

Above: An apse or "Bed-womb" creates a safe sleeping space.

The thick Superadobe walls create shade, and their thermal mass is cooling in the daytime and warming at night. In hot climates the roofless dome acts like a chimney funneling hot air upwards and drawing in cool air from adjacent spaces.

Some uses for a roofless dome design include:
1) A community kitchen such as those found in some African and Indian villages amongst others, with an open central fireplace, and attached storage apses.
2) A safe pen to keep livestock at night where the owner or herder can watch over them from the "bed-womb"/apse.
3) A secure area for drying and storing agricultural products, where the roof frame and netting can support dried produce spread over the top surface in the sun, or hanging down from the frame in the dry shade.
4) A sheltered area for growing delicate food crops in harsh climates such as dry, hot deserts with wind-blown sand, or very cold deserts with freezing winds.

Above: Finishing the exterior with Reptile plaster.

Left: Finished Reptile exterior with white grout, and a planter.

Below: The Roofless Dome as part of the village courtyard.

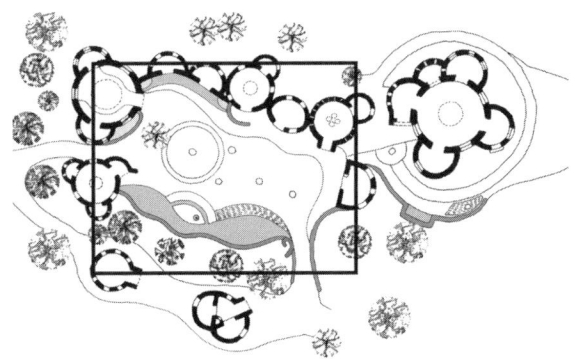

COURTYARD
AND LANDSCAPE

View of the courtyard in winter.

SUPERADOBE - SANDBAG SHELTER

Right: Landscaping and short retaining walls leading to the sloped entry path, for the handicapped, and the drainage area.

THE COURTYARD: Landscaping.

The entire courtyard has been dug out two feet and its earth used to build the shelter structures surrounding the yard, as well as its landscaping.

The natural grade is contained by low retaining walls along one side of the courtyard. Superadobe flexibility makes it easy to follow the contour of the land. Traditional hillside agriculture in many countries used short retaining walls to "terrace" the sloped landscape like giant steps. Gravel behind the retaining walls and weep holes are usually added to relieve the water pressure in the rainy season.

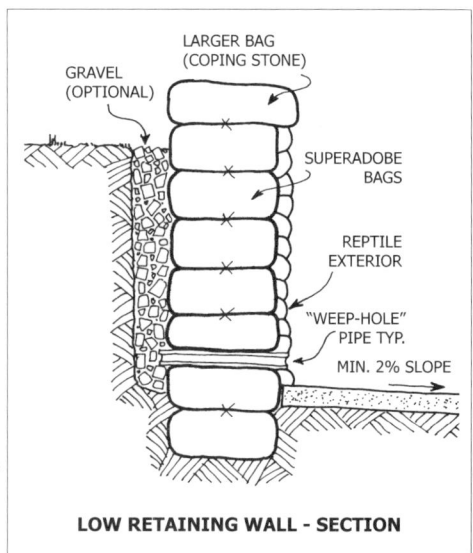

LOW RETAINING WALL - SECTION

Right: Low retaining walls can "terrace" an existing hillside to assist planting which will stabilize the slopes.

Left: Filling several layers of Superadobe for the retaining walls, where the grade is sloped to form a ramped entrance into the courtyard, facilitating disabled access.

Below: Completed retaining wall between two domes.

Above: The coils turn tightly around a drinking fountain, seen here in the winter snow.

By carving down into the ground, the flexible layers of retaining walls can also become landscaping elements such as seating, a planter, a pond, stepped amphitheater seating, sculptural lighting, a drinking fountain and so on. The Superadobe coils, which so easily reflect the wave motion of water are combined here with a drinking fountain and a pond, two elements which bring water into the community space of the courtyard.

The sunken courtyard, when properly sloped to a drainage system, will guide rainwater to the trees, as well as channeling flood water away from the structures.

Drinking Fountain

On every continent the traditional village well was also a meeting place where residents would go to draw water, and where conversations, visions, and moments of poetry took place. Here, a drinking fountain protected by tightly curving Superadobe coils becomes a focus and a sculptural element, as well as centralizing the fountain, water hose, and electrical outlets. Although water and power are not usually available in emergency situations, for a permanent eco-village these can be supplied by catchment water, pumps, and solar or wind power in remote areas. Endless sculptural possibilities exist for creating such elements of functional, poetic and social focus.

Left: The drinking fountain, a water faucet with hose pipe, and electrical outlets are part of the coiled sculpture, combining the utilities.

Below: The sculptural drinking fountain utility is a celebrated social focus.

SUPERADOBE - SANDBAG SHELTER

Above: Shaping the bags with a brick.

Right: Painting the waterseal onto the pond before it is filled with water. The fountain's hose-jets have been encased in reptiles, stabilized earth mud-balls.

Pond with Fountain

A pond with a fountain can also be constructed with the Superadobe coils. This not only has the poetic identity of regeneration, but also serves a practical purpose of collecting water from the courtyard and, like the well, making it available in the dry months. Nature's cycle of filtering the rainwater through the earth, regenerating it deep underground, and bringing it forth as a spring of fresh water, can be repeated here on a small scale. Depending on its size and design, a pond can contain fish and vegetation. The entire landscape, integrated with the structures, can be planned as a Permaculture eco-system.

Left: Pond, fountain, retaining wall, seating and sleeping platform.

Below: Stabilized Superadobe forms the edge of a larger lake. The Hesperia Lake project was funded by FEMA, and the Superadobe coils continue to protect the shoreline from erosion over a decade later.

The pond shown here creates a cooling element for the courtyard in the summer. The bag construction is plastered with cement/lime stabilized earth and sealed with a nontoxic waterproof paint on the finished plaster surface. Larger ponds can be lined with plastic sheeting liners before plastering, or use traditional earthen mixes, such as the traditional Iranian "Sarooj" mix (clay, lime, ash and sand).

The curved seating, as a sleeping surface for the summer, is constructed by small retaining walls and is packed from behind with well drained earth. The seating surface is sloped towards the courtyard for drainage, and finished with a paving of bricks set into cement stabilized mortar and painted with a sealant, if needed.

Right: A mother and children cool off in the pond in summer.

Below: The ripples of the fountain enliven the pond and send the sound of water throughout the courtyard.

Rainwater Catchment

Many traditional courtyard designs in countries with warm climates use the courtyard floor to catch the fresh rainwater and channel it into underground cisterns or wells. There it stays cool until it is used in the dry, summer season. Today we can use solar powered pumps to pull up the water for agricultural gardens, fountains, washing, cooling and, under the right circumstances, for drinking.

Right: A tree grows from the water catchment drain, surrounded by seating in a welcoming curve. The rough surface of the perforated brick paving prevents slipping on the ramp, for pedestrians as well as wheelchairs.

Left: The drainage pit is filled with recycled materials - broken rocks, bricks and several pipe sections topped with perforated bricks, to quickly drain the water below grade.

Below: Finish paving of dry set perforated bricks is added over a mesh and/or gravel bed.

This courtyard simply catches and channels the water underground via a "dry well", which is placed towards the trees and landscaping. The water is not collected into a cistern or underground chamber, but is quickly drained below the surface through recycled pipe sections and a pit of broken bricks and rocks. If needed, this can be lined with plastic or fabric sheeting to prevent it silting up. The finished paving must allow the water to percolate through to the dry well. This is a basic type of dry well and is suitable for well drained soils. In heavy clay soils a standard drain pipe or french drain with gravel is needed to move the water away from the courtyard.

Above and below: Patterns and walkways are designed on site, using a straight edge to level the ground and set the brick patterns at a slope for drainage.

Right: Sections of the floor are poured and trowelled up to the brick level.

The Courtyard Floor

The paving for the courtyard combines small units such as bricks, pavers, or stones, and a poured surface such as a stabilized earth or concrete slab.

Left and below: The edge sections are poured and trowelled using the bricks to level the wet mix.

The following methods can be used for paving as well as seating surfaces, using brick pavers and a stabilized earth or concrete slab.
1) The ground is levelled and compacted. For the floor of the courtyard, the earth is levelled using a long straight edge and compacted with a plumber's tamper.
2) It is sloped towards the drainage area. The floor is given a minimum 2% slope towards the dry well under the trees.
3) Brick patterns are arranged to a consistent height. The whole courtyard floor is divided by brick patterns to provide expansion joints and minimize cracking.

Right: Giving the surface a broom finish to prevent slipping.

4) The slab is poured with or without steel mesh, depending on its size. By combining brick patterns with the slab, smaller floor areas can be poured at one time. This allows the builder more flexibility with material, manpower, and design.
5) The surface is finished with a steel trowel or "float". The bricks determine the level or thickness of the finished floor, avoiding the need for special markers.
6) The surface is brushed before it sets to give a nonslip finish. Expansion joints can be scored while the slab is still wet.
7) If needed, the surface is sealed.

Left: Scoring the slab for expansion joints before it sets.

Below: Finished floor with brick patterns and expansion joints.

Above and below right: The courtyard is used for community activities of all ages.

8) Even after the floor slab is poured the brick areas could be removed to allow space for planting, walls, or more structures. In time, the divided areas, expansion joints and bricks, allow the floor to settle or expand with the climate and earth below.

216   KHALILI - CAL-EARTH

Above: The courtyard seating is widened to provide sleeping areas. The paver bricks are widely spaced to harmonize the curved and straight edges.

Any physical imperfections in the materials and workmanship become unimportant when the eye is attracted by the organic geometry of the entire courtyard landscape. Organic geometry empowers an unskilled builder to manifest his or her intuitive designs.

Left: The courtyard floor as it relates to the domes and pouches.

SUPERADOBE - SANDBAG SHELTER

The Courtyard floor and dance circle.

*"humans are members of the same body
all created from the same essence
if one limb is in pain
others cannot be at ease
if you are indifferent to other people's pain
you cannot be called a human"*

- Saadi
(13th century Persian poet)

C·H·A·P·T·E·R **3**

U.N., NGO, AND PRIVATE SHELTER PROJECT EXAMPLES

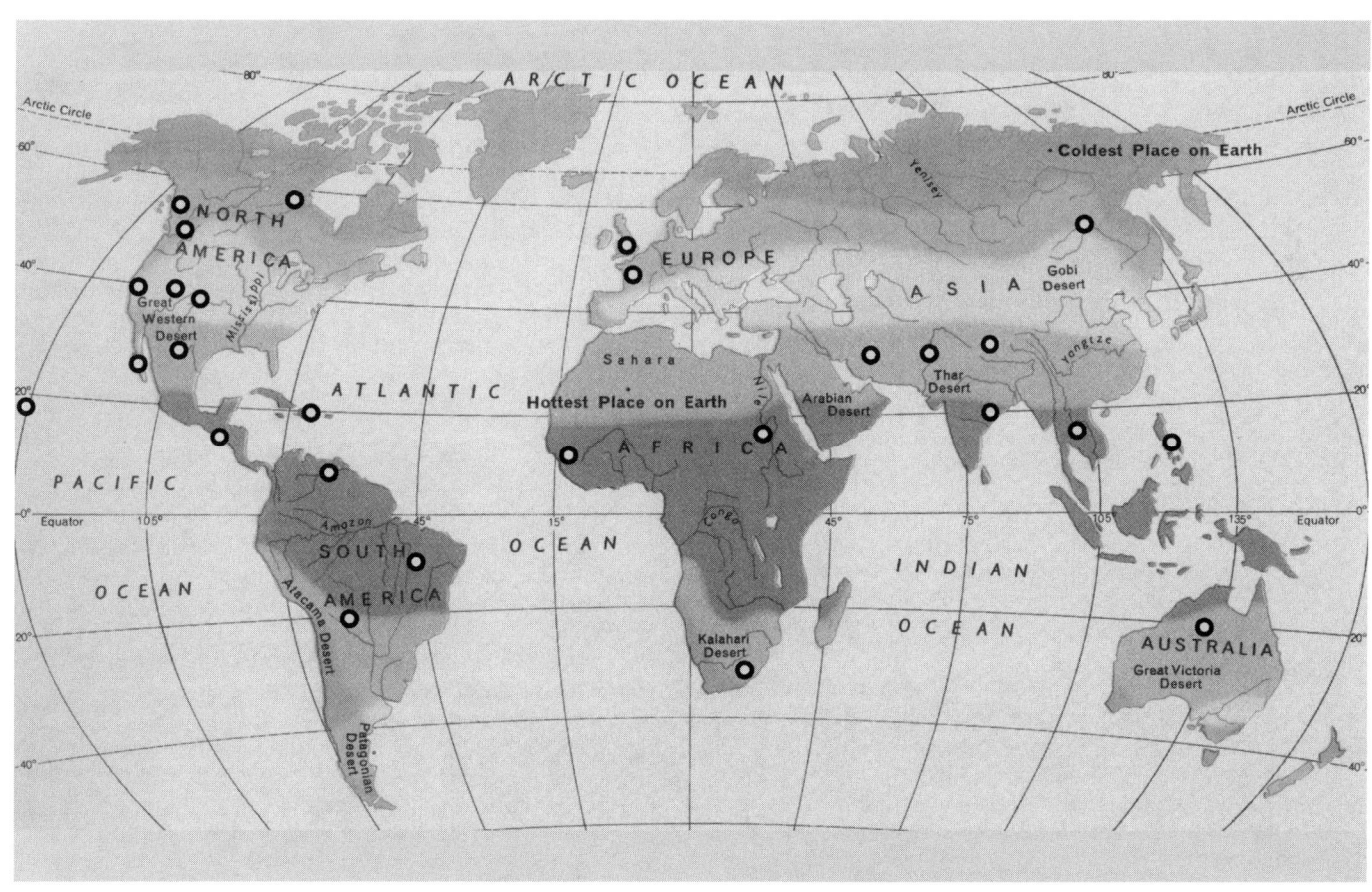

○ Areas in the world where people are already turning the materials of war, sandbags and barbed wire, into the building blocks of peace.

Relief Resources

30 Jul 2001............Sandbag homes may be shelter breakthrough... ....By Anton Ferreira

UNITED NATIONS (Reuters) - Senior U.N. officials plan to test a building method using sandbags and barbed wire they say could revolutionise the way emergency housing is provided after natural disasters such as floods, earthquakes and hurricanes.

The officials told Reuters that the method, known as "Superadobe" and developed in Hesperia, California, by Iranian-born architect Nader Khalili, could provide durable, cheap shelter very quickly after calamities like the Gujarat earthquake earlier this year in India.

"I thought it was amazing. It is a hidden treasure," said Omar Bakhet, director of the Emergency Response Division at the U.N. Development Program.

Bakhet and his program adviser Lorenzo Jimenez de Luis visited Khalili's California research site earlier this month and said they immediately realised the potential of his building method.

"The technology is fascinating," Bakhet said. "It's a technique one can learn in a few days."

The Superadobe method involves filling empty sacks with earth dug from the building site and piling them in layers with strands of barbed wire acting like Velcro to provide added stability.

The simplest design is a circular room tapering toward the top to form a dome that sheds snow or rain. Several examples of the beehive-like structures have been built in Hesperia and elsewhere, and they have passed seismic testing required under California's strict earthquake-zone building codes.

Building with Superadobe requires no special skills, and rooms can be added. Khalili has spent most of his career designing affordable housing for the homeless, but until now his work has had little attention from disaster relief professionals.

"I don't think there's any risk, it's a proven technology," said Bakhet. "It's cost effective, you need very little building material, just what nature gives you." Bakhet and Jimenez de Luis said the only problem they foresaw was persuading governments to try the new technology.

"If these structures had the shape of a conventional house, it would be much easier," said Jimenez de Luis. "A government is going to be reluctant to accept a hemispherical thing." He said Superadobe represented a far better option than the tents or plastic sheets and corrugated iron that are used now to provide emergency shelter for refugees from natural disasters or wars.

The Superadobe technique can be learnt in a few days.
Photo: California Institute of Earth Art and Architecture

ABSOLUTELY PERFECT "The (Khalili) initiative is very suitable because it covers the permanent character of the structure and the dignity aspect of the people who are going to benefit from the shelter - to live in one of these houses is absolutely perfect. To live in a tent is not so dignified in the long run."

Jimenez de Luis said Superadobe structures would also be better able to withstand future earthquakes or floods. This was important in regions like Central America or the subcontinent that experienced recurrent disasters.

"The (concept) is extraordinarily positive and definitely worth testing," he said. "It's just a matter of trying it once or twice for this thing to fly solo."

The U.N. officials said they were determined to launch a pilot project, possibly in Gujarat where some 1 million homes are needed, if the backing of local authorities could be obtained.

"Here you have a technology that's so simple, so effective and can be used by everybody, you are cutting the time for addressing housing needs by I don't know how many percent," Bakhet said. "But like all new approaches, how many people would be readily prepared to embrace it? We are all afraid of anything new... So this where the challenge is."

Khalili said in a telephone interview from Hesperia that he dreamed of building an entire city in India.

"I showed them the plans I have for houses, clusters from 1,000 to 5,000 to 10,000, all the way to a million-person town that will be totally sustainable... "Imagine, if they gave me 1,000 soldiers and a couple of hundred students, I could build a whole town for them... If you can cut through the bureaucracy, I have the design," he said.

The above article is reprinted with permission of Reuters Alertnet www.alertnet.org

Above and below: The visiting UNDP team, participate in constructing a Superadobe shelter.

United Nations (UNDP, UNHCR, UNIDO, UNITAR)

During the construction of the Seashell Dome, Cal-Earth was visited by the director of Emergency Reponse and his team from the United Nations Development Program (UNDP) to assess the Superadobe system for emergency shelters around the world. They learned the technology hands-on using bags, barbed wire, and cans of earth. By the end of the day they were impressed enough with the speed and ease of construction to leave their hotel and spend the night in the shelters to test their qualities. Thus the prototype "United Nations Village" was named. A few days later, from the United Nations' headquarters in New York they gave their opinion in an interview to Reuters News agency.

The qualities they experienced were the sustainability of these timeless forms and materials inherent in the universal elements, and within every human; the same forms in nature that design the human skull and the space shuttle. They saw that ancient and new forms are one and the same, from historical buildings, water reservoirs and wind-scoops to space shuttles and nuclear reactors; that a shell structure can build an eggshell or the fuselage of an airplane, from planetary colonies to a simple coiled emergency shelter.

Above: The visiting UNDP team, Director Omar Bakhet, Lorenzo Jimenez de Luis, with their colleague Afsaneh Bassir-Pour and Manijeh Mirdamad, with Hesperia's Mayor, Nader Khalili and the Cal-Earth team.

Visiting three times from UNITAR (United Nations Institute for Training and Research), Nassrine Azimi, then director of the New York office, initiated the "Training the Trainers" proposal after Hurricane Mitch in Central America.

The sandbag shelter technology met the stringent requirements of UNDP and the United Nations High Commission for Refugees (UNHCR) for emergency shelter.

In 1994 it was chosen for the Baninajar refugee camp pilot project in Khuzestan province, Iran. The technology and design were selected for the following characteristics:

1) Minimum size (varying from 45 to 200 sq. ft.); minimum shelter is specified according to United Nations requirements.
2) Insulation against cold and heat far superior to tents.
3) Minimum cost (U.N. costs were $621 per shelter for 200 sq. ft.).
4) Maximum speed (each shelter can be built in one to seven days depending on size and soil stabilization)
5) Minimum on-site skill (refugees with no experience can build under the supervision of a trained builder).
6) Temporary with the potential for permanence (by using stabilized or unstabilized earth, with or without protective exterior finishes); host countries often do not allow permanent homes, to encourage refugees to return to their homeland.

Most refugee camps are designed as grids with no thought of social structure leaving the refugees disoriented and creating social instability. By contrast, teaching the refugees themselves how to build with Superadobe, empowers them to participate in building their own family shelters. The flexibility of the Superadobe for design, allows a community to incorporate the social and spatial structures needed for stability and a sense of continuity in shattered lives.

Baninajar Refugee Camp

In early 1995, after building two prototypes for testing at Iran's Ministry of Housing Research Center, UNDP in collaboration with UNHCR built fourteen refugee shelters at Baninajar camp in the south of Iran, after the Iraq-Kuwait war.

The 5 meter / 15 ft. diameter domes were built by the refugees themselves under the supervision of U.N. architect Hamid Irani, who had been trained at Cal-Earth in their construction. They were constructed in a three week period, averaging one shelter every six days. By lifting only one small can of earth at a time (not a shovel) and by filling the bags directly on the wall (no lifting heavy weights), the whole family including women, children and the elderly were able to participate. In contrast to concrete block/corrugated sheeting and tents, these shelters were cool in the hot desert. They used local materials and were later upgraded with finishes, to be inhabited for at least two years before the refugee camp was finallly dismantled.

The United Nations report which follows noted the details and success of the work showing that the shelters did meet the requirements of UNHCR, the host country and the refugees. Their recommendations were incorporated into later projects.

Above and below : Shelters at Baninajar refugee camp.

SUPERADOBE - SANDBAG SHELTER

United Nations Development Programme (UNDP)
in cooperation with
United Nations High Commissioner for refugees (UNHCR)

" Sandbag Shelters" Pilot Project (Phase II)
Construction of 14 Units Shelters at Baninajar Camp.

Report prepared after by:
Yasmin Mazhari, Mohammad Hassan Dashti (Technical Assessment), the independant team for UNDP, Iran.

Excerpts from the report of Sandbag Shelters Pilot Project after one and a half years of use: August 1996

6. Recommendations

6.1. Social Acceptance

Social and cultural acceptance of this form of housing is one of the primary factors to take into careful consideration before undertaking large-scale construction. This will help in assessing the suitability of the shelters to the intended case-load and in determining the most appropriate design. For example, the beneficiaries of this project were unaccustomed to a communal living space for both sexes. Prior consideration of this fact would have shown that two smaller connecting domes would have been suited to their needs.

6.2. Emergency Situations

The simplicity and strength of design, speed and ease of construction are the key elements of Sandbag Shelters which ideally should make them a feasible alternative to traditional emergency shelters. However, this will remain purely an assumption until construction has been tested in a real emergency situation, against veritable time and logistical constraints.
Sandbag Shelters should not be solely limited to an option for refugee housing. Due to their strength in withstanding earthquakes and heavy winds, they could also be applied to provide emergency shelter for peoples displaced by natural disasters or even as permanent housing for those living in disaster prone areas.
Should Sandbag Shelters be accepted as a viable option for emergency shelters, it would be worthwhile to provide training in their construction to workers from national emergency task forces. In this way, relief workers with access to disaster struck areas would be in a position to immediately commence construction of shelters, and to act in the capacity of supervisors in instances where beneficiaries are involved in the construction of their own shelters. This foresight will help save crucial time in an emergency situation.

6.3. More Technical Recommendations

Spatial arrangement. When a number of buildings / shelters are placed next to one another, there should be harmony in the arrangement, even in a chaotic situation of a disaster emergency.
We, the writers of this Report, feel confident now that "Sandbag Shelters" work. And with good on-site management, strong cost-control and attending to the cultural and technical recommendations given in this Report, it may succeed as a strategy for housing refugees in Iran. To account for contingency before it arises, would guarantee this success .
Furthermore, it is vital to respect the culture of beneficiaries in the various projects. A commitment must be made to speak with the beneficiaries, ask their views and win their confidence before the project starts.
Finally, it would be highly recommended, that, given the present inflation in the country, jute (and/or plastic) bags are purchased in big quantity and stocked for future projects as they account for 50% of the total cost of each shelter. This will ensure cheaper prices and also guaranteed quality. Vendors can prepare the bags in the exact measurements that may be specified later. According to the designer, Mr. N. Khalili, it is advantageous to have bags in longer lengths so as to reduce the number of joints and thereby strengthen in structure.
We recommended necessary modifications in the design, should sandbag shelters be prescribed for cold regions. The form in which this project was designed and executed is rather more compatible with hot and dry climate. The relatively high ceiling of shelters is good for circulation of warm air to the top and cold air to the lower areas. The dome shape of ceiling receives sun's rays only from one side, and thus the other side is more cool.

(Excerpts from the report of Sandbag Shelters Pilot Project continued)

Annex I- Sandbag Shelters - Actual Cost of Materials, Tools & Labor

Phase I: The Emergency Shelter ("Family Kit")

Items	Quantity	Cost IRR	Total Cost IRR	Total Cost USD
Materials:				
Sandbags	1,500 units	620 each	930,000	310
Barbed Wire	250 Kg	2500/Kg	625,000	208
Nails	25 Kg	2500/Kg	62,500	21
Metal Wire	50 Kg	2200/Kg	110,000	37
Plastic Sheeting	100 sq.m.	4500/sq.m.	45,000	15
Tools:				
Shovels	2	7,500	15,000	5
Pick	1	5,000	5,000	1.5
Wire Cutter	1	9,000	9,000	3
Level	1	5,000	5,000	1.5
Measuring Tape	1	5,000	5,000	1.5
Gloves	10	2,500	25,000	8
Cutter & blades	1	5,000	5,000	1.5
Rope	10m	300/m	3,000	1
Metal Pipe (5m long)	1	21000	21,000	7
Total Cost of Emergency Shelter:			1,865,500	$621

Phase II : Upgraded Shelter (includes all items from Phase I plus following options)

Items	Quantity	Cost IRR	Total Cost IRR	Total Cost USD
Materials:				
Clay Soil	35 tons	2,000/ton	70,000	23
Straw	500 Kg	200/Kg	100,000	33
Gypsum (50 Kg bags)	40 bags	2,500/bag	100,000	33
Door (Highest/Handle/Lock)	1	140,000	140,000	47
Windows	4	80,000 - 1set	320,000	107
Flooring (concrete)		Lump sum	100,000	3
Electricity (Light/Wiring/Switch/Outlets)		Lump sum	100,000	3
Windcatcher		100,000 each	100,000	33
Fireplace		Lump sum	200,000	67
Cost of Upgrading Options:			1,230,000	$409
Cost of Upgraded Shelter:			3,095,500	$1,030
Cost of Labor:				
Basic Shelter			500,000	167
Interior Coating			120,000	40
Exterior Coating			180,000	60
Site Work			30,000	10
Total Labor Cost per Shelter			830,000	$277

*Exchange Rate: IRR 3,000 = USD 1

Information and photos are arriving on the network from around the globe of sandbag - Superadobe shelters built by the Cal-Earth alumni, examples of which are shown here.

Global Shelters and Eco-Domes

People are learning to build with sandbag shelters all over the world from small domes to multi-dome and vault houses, for orphanages, clinics, schools, and most of all prototypes to demonstrate and teach. By continuously training the trainers who return to their homelands to build and teach, or who travel afar travel to help others, Cal-Earth journeymen and journeywomen have trained individuals and NGOs by building with Superadobe.

The largest of these organized relief efforts was in northern Pakistan after the severe Central Asian earthquake in 2005, as a collaboration between Cal-Earth and the SASI foundation. According to SASI, hudreds of refugees were trained hands-on, as well as national and international NGOs and military emergency relief units. The needed

rolls of bags were distributed to the refugees. Since the structural system passed tests in California, the shelters were approved by the authorities in Pakistan. Costs for the training shelters were similar to the UNHCR cost some years earlier.

Pictures keep arriving, and so many more are being built around the world of which we have no data. Our thanks goes to all the Cal-Earth apprentices, teachers and journeymen, many of whom we acknowledge in the first pages of this book. To do justice to their contribution, Cal-Earth is planning a future book on their work to introduce them and their projects.

SUPERADOBE - SANDBAG SHELTER

Cal-Earth Pakistan

Cal Earth, Pakistan has been formed to respond to the largest crisis Pakistan has ever faced.
The October 8th 2005, Earth Quake

Sasi Foundation has joined hands with **Cal-Earth Foundation, USA**
based on the Cal-Earth "SUPERADOBE" technology.
It has undertaken the largest community based participatory reconstruction drive ever
where all Pakistanis and the world is coming together.

It currently is training the Pakistan Army, Azad Jammu Kashmir officials and NWFP government officials.
Besides it is developing its own team and training NGO's interested.
It has started its first training session at the Army Gymnasium Compound in front
of the Pakistan Army, General Headquarters (GHQ) off Mall Road, Rawalpindi, Pakistan.

Second Cal Earth Pakistan e shelter training site in Islamabad started
under the patronage of Ministry of Interior SWS, walking distance to Serena Hotel, Islamabad, Pakistan.
Cal-Earth Pakistan e shelter training site started in collaboration with
Institute of Architect Planners (IAP) at F-9 Park, Islamabad.

Contact: Shahid Malik Shahid,   Khurram Shroff

www.calearthpakistan.org

A Model Summit: Showing effective leadership in crisis
MUSHARRAF LEADS PAKISTAN'S DISASTER MANAGEMENT

".....The Pakistan army has already approved and is beginning to permit construction of Eco-Domes by Cal Earth Pakistan, a joint venture between California-based Cal Earth and Pakistani American entrepreneurs. USAID, the US government's partner in disaster relief, has committed to provide funds for low-cost housing solutions. Rather than buying more tents than can be deployed or are needed, USAID's able management team, which will be in Islamabad for the conference, should look seriously at investing its resources in these earthen super-adobe style homes........."

National Review: Friday, November 18, 2005.

"every man and woman is a doctor and a builder to heal and shelter themselves"

Photos from Cal-Earth Pakistan training refugees to build shelters after the 2005 Central Asian earthquake.

SUPERADOBE - SANDBAG SHELTER

The Aga Khan Award for Architecture 2004

The Ninth Award Cycle, 2002-2004

Sandbag Shelter Prototypes
various locations

Architect   Cal-Earth Institute, Nader Khalili, US
Timetable   First development, 1992

Description

The global need for housing includes millions refugees and displaced persons – victims of natural disasters and wars. Iranian architect Nader Khalili believes that this need can be addressed only by using the potential of earth construction.

After extensive research into vernacular earth building methods in Iran, followed by detailed prototyping, he has developed the sandbag or 'superadobe' system. The basic construction technique involves filling sandbags with earth and laying them in courses in a circular plan. The circular courses are corbelled near the top to form a dome. Barbed wire is laid between courses to prevent the sandbags from shifting and to provide earthquake resistance. Hence the materials of war – sandbags and barbed wire – are used for peaceful ends, integrating traditional earth architecture with contemporary global safety requirements.

The system employs the timeless forms of arches, domes and vaults to create single and double-curvature shell structures that are both strong and aesthetically pleasing. While these load-bearing or compression forms refer to the ancient mudbrick architecture of the Middle East, the use of barbed wire as a tensile element alludes to the portable tensile structures of nomadic cultures. The result is an extremely safe structure. The addition of barbed wire to the compression structures creates earthquake resistance; the aerodynamic form resists hurricanes; the use of sandbags aids flood resistance; and the earth itself provides insulation and fireproofing.

Several design prototypes of domes and vaults were built and tested. The system is particularly suitable for providing temporary shelter because it is cheap and allows buildings to be quickly erected by hand by the occupants themselves with a minimum of training. The shelters focus on the economic empowerment of people by participation in the creation of their own homes and communities.

Each shelter comprises one major domed space with some ancillary spaces for cooking and sanitary services. Incremental additions such as ovens and animal shelters can also be made to provide a more permanent status and the technology can also be used for both buildings and infrastructure such as roads, kerbs, retaining walls and landscaping elements.

Because the structures use local resources – on-site earth and human hands – they are entirely sustainable. Men and women, old and young, can build since the maximum weight lifted is an earth-filled can to pour into the bags. Barbed wire and sandbags are supplied locally, and the stabilizer is also usually locally sourced.

Since 1982, Nader Khalili has developed and tested the Superadobe prototype in California. In 1991 he founded the California Institute of Earth Art and Architecture (Cal-Earth), a non-profit research and educational organization that covers everything from construction on the moon and on Mars to housing design and development for the world's homeless for the United Nations. Cal-Earth has focused on researching, developing and teaching the technologies of Superadobe. The prototypes have not only received California building permits but have also met the requirements of the United Nations High Commissioner for Refugees (UNHCR) for emergency housing. Both the UNHCR and the United Nations Development Programme have chosen to apply the system, which they used in 1995 to provide temporary shelters for a flood of refugees coming into Iran from Iraq.

Khalili's educational philosophy has also continued to develop. A distance-teaching programme is being tested for the live broadcast of hands-on instruction directly from Cal-Earth. Many individuals have been trained at Cal-Earth to build with these techniques and are carrying this knowledge to those in need in many countries of the world, from Mongolia to Mexico, India to the United States, and Iran, Brazil, Siberia, Chile and South Africa.

In 2004 the Aga Khan Award for architecture went to "Sandbag Shelter Prototypes", from their research and development to the prototype application in different countries. This Description, Jury Citation, and Project Data, describe the qualities for which the award was given.

Jury Citation

These shelters serve as a prototype for temporary housing using extremely inexpensive means to provide safe homes that can be built quickly and have the high insulation values necessary in arid climates. Their curved form was devised in response to seismic conditions, ingeniously using sand or earth as raw materials, since their flexibility allows the construction of single- and double-curvature compression shells that can withstand lateral seismic forces.

The prototype is a symbiosis of tradition and technology. It employs vernacular forms, integrating load-bearing and tensile structures, but provides a remarkable degree of strength and durability for this type of construction, which is traditionally weak and fragile, through a composite system of sandbags and barbed wire. Created by packing local earth into bags, which are then stacked vertically, the structures are not external systems applied to a territory, but instead grow out of their context, recycling available resources for the provision of housing. The sustainability of this approach is further strengthened because the construction of the sandbag shelters does not require external intervention but can be built by the occupants themselves with minimal training. The system is also highly flexible: the scale of structures and arrangement of clusters can be varied and applied to different ecosystems to produce settlements that are suitable for different numbers of individuals or groups with differing social needs. Due to their strength, the shelters can also be made into permanent housing, transforming the outcome of natural disasters into new opportunities.

Project Data

Architect   Cal-Earth Institute, US: Nader Khalili, concept and design; Iliona Outram, Project Manager.

Consultants   P. J. Vittore Ltd, US, and C. W. Howe Associates, US, structural engineers.

Sponsors and clients   National Endowment for the Arts, US; Southern California Institute of Architecture (Sci Arc), US; the Ted Turner Foundation, US; United Nations Development Programme (UNDP), US and Switzerland; United Nations High Commissioner for Refugees (UNHCR), Iran offices; the Bureau for Alien and Foreign Immigrant Affairs (BAFIA), Iran; Laura Huxley's Our Ultimate Investment Foundation, US; the Rex Foundation, US; Kit Tremaine, US; the Leventis Foundation, Cyprus; the Flora Family Foundation, US.

Prototypes built to date by   Hamid Irani and Iraqi refugees at Baninajar Camp, Iran; Eric Hansen, Mexico; Djalal and Shahla Sherafat, Canada; Michelle Queyroy and orphans at the MEG Foundation, India; Dada Krpasundarananda, India, Thailand and Siberia; Mara Cranic, Baja, Mexico; Virginia Sanchis, Brasil; Patricio Calderon, Chile; Jim Guerra and Mexican farmworkers, US; Don Graber, Craig Cranic, Giovanni Panza and Yacqui People of Sarmiento, Mexico.

Timetable   Sandbag Shelter Prototypes (Superadobe): first development, 1992.

The following Cal-Earth alumni sent these photos of their projects:

Claire Blanchemanche
Bradley Cebeci
Mara Cranic, Craig Cranic
Neil Decker
Hooman Fazly
Anton Ferreira
Don Graber
Jim Guerra
Eric Hansen
Mark Harmon
Martin Hartman
Hamid Irani
Michael Huskey
Nadia Janjuan
Dada Krpsundarananda
Ian Lodge
Alessandro Lopezy
Betty Marvin
Theodore Petroulas
Pegasus Project
Virginia Sanchis
Djalal and Shahla Sherafat
Ulises Ramirez

"your imagination my friend flies away
then pulls you as a follower
surpass the imagination and
like fate itself
arrive ahead of yourself"

- Rumi

C·H·A·P·T·E·R 4

CHILDREN
AND THE FUTURE

"And today, experiencing this sutainable earth architecture, are the children - the builders of the future."

SUPERADOBE - SANDBAG SHELTER

Above: The individual elements of the village cluster around the courtyard to make a sustainable community vision.

Below: Child emerges from an earth shelter.

The Vision, Children, and Superadobe

In conclusion, we can physically touch Rumi's poem: "earth turns to gold in the hands of the wise." The structures for an Eco-Village and Eco-Domes have been created by using knowledge, earth, and locally available materials: sandbags, barbed wire and a few additional materials such as a stabilizer, shovels, and compass chains.

"The Vision" of this sustainable earth architecture, is building with the "Timeless Principles" of arches, domes, vaults, and organic forms, and the "Timeless Materials" of the universal elements, earth, water, air and fire in the spirit of unity. Any individual or community can not only dream of an Eco-Village, but participate hands-on by building it with the most abundant material under our feet: Earth.

But an Eco-Home and Eco-Village does not begin or end with construction and hands-on community participation alone.

"Sustainability begins with a person; if you can sustain your own ideals and quests, you are the best member of a sustainable society," we often say to groups of visitors and workshops.

Now the children, by distilling the learning, and experiencing this architecture through all their senses, become the sustainable builders of the future. It all starts with children, "our ultimate investment" according to Laura Huxley.

Structures built to the human scale are irresistible to climbing by children and adults.

Left: Children find a seat on the top of a wind-scoop.

Below: The lacy brick patterns of the "Rumi Dome" invite all to climb.

SUPERADOBE - SANDBAG SHELTER

Right: The Village clustered plan can now be seen at a human scale as a theater celebrating human events.

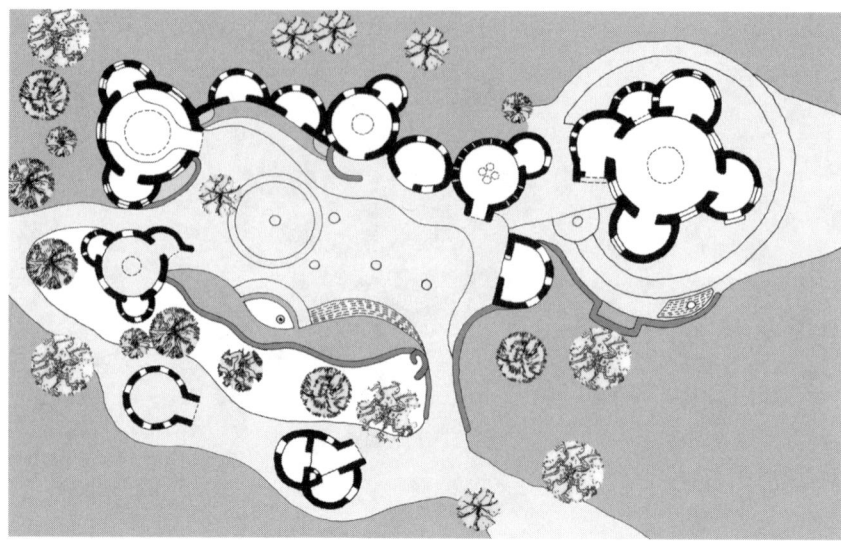

Above: The earth transmits its anwers with all five senses at once.

Right: The dancers express life in an Eco-Village, carved out of the earth.

The Future

Where do we go from here, you might ask? And after these pages of how-to build sandbag shelters you may have many more ideas and questions than when you started to read. After years of research and development, education and explanation to all who came our way, we must admit that for us too the questions keep on coming.

Yet we have found that the most puzzling and fundamental questions were usually answered when we stopped talking and started digging. When we put our hands in the earth it transmitted its answers, and the process of hands-on work, learning with all five senses at the same time, gave the most useful knowledge of all for human habitats.

This process of building spreading around the globe has made it ever harder to complete the entire manual on Sandbag Architecture and resulted in it being published one volume at a time.

Left: It all begins with the children and their eager questions, who are poised to enter a sustainable future.

In completing this volume it has been hard to limit ourselves to communicating just the essentials of building with Superadobe, sandbags and barbed wire, or earth-bags. The greatest contribution to this process has been from children, who by their eager questions have molded the response - simplify, simplify, simplify.

Naturally, the next stage of this work is to build a small permanent house that will be called "Eco-Dome". This house is also called "Moon Cocoon" as it takes the past and moves it into the future. It is sustainable, cooled by the wind, heated by the sun, a non-toxic environment, yet carved out of the earth on the site; a home for the body and soul.

"It takes students to draw out the milk of knowledge from the teacher"

- Rumi

"It is all a journey"
- Rumi

Left: The Eco-Dome seen at a distance through the Rumi Dome.

Eco-Dome Plan

Right: Walking towards the Eco-Dome from the village plaza.

Below: The Eco-Dome entrance.

MODEL 8: Eco-Dome (Moon Cocoon).

After learning to build basic sandbag shelters you are ready to start permanent Superadobe structures using advanced techniques, which include the amenities and finishes we come to expect today.

One such home is called "Eco-Dome (Moon Cocoon)". Its clover leaf shaped plan of 400 sq. ft. (40 sq. m.) includes a living room, bed-womb, kitchen, and bathroom. The latest approved blueprint and engineering is for a double dome of approx. 800 sq. ft. (80 sq. m.) It uses the same materials as the shelters and can make an exquisite small earth home.

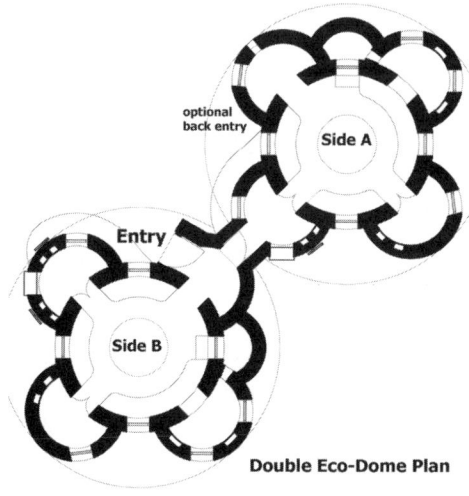

Double Eco-Dome Plan

Above: Interior of the "bed-womb" niche, looking from the main dome.

Left: Interior of the main dome looking towards the entry apse - an earth home built with non-toxic, natural materials. The right hand window is the windscoop.

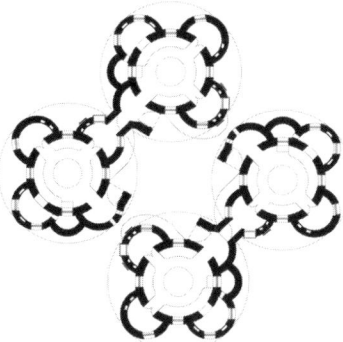

Eco-Dome Courtyard Cluster

The earth used to build it is carved from around the structure by creating landscape. As a low-cost, self-help home, to be built by 3-5 people, it is a natural sculpture in the color and form of the earth itself. The construction documents for this house (plans, elevations, sections, energy calculations, and engineering), have been approved and built in California, with its strict seismic zone 4 codes. A cluster of these Eco-Domes can fulfill the vision of an Eco-Village.

We hope and expect that the next volume about the Eco-Dome will bring the inspiration and knowledge, to keep working hands-on with this limitless flexible building technology.

SUPERADOBE - SANDBAG SHELTER

Above: The Eco-Dome, moon cocoon, is the last in line at the end of the village. It is a cluster of apses/pouches around a dome, growing from the earth itself, cooled by the wind, warmed by the sun.

Distance Learning and the Future

As the human family grows and spreads all over the globe, enabled by tomorrow's communication technology, it is inevitable that the students will become the teachers of tomorrow, and their tools are distance learning, eco-communities, and non-toxic, natural materials and more. Ultimately this will result in a sustainable environment brought about by the equilibrium of the universal elements of earth, water, air and fire.

Cal-Earth is hoping to participate in this exciting and dynamic growth in empowering people not only to build for themselves, but to contribute their talent to the ever increasing need for human shelter. It has now a library of data, books, films, and kits that are being constantly developed and expanded to teach people wherever they are, and under whatever conditions.

The focus is to build structures not only to respond to immediate needs in a disaster, but also to ensure their sustainability into the future as permanent living and working environments.

"timeless materials, timeless principles"

APPENDIX I

STRUCTURAL PRINCIPLES

P. J. VITTORE, Ltd.
Arlington Heights, Illinois

The Structural Integrity of Domes and Vaults

The engineering community as well as most governmental engineering review groups need to understand the basic differences in engineering domes and vaults versus engineering design of beam and column buildings. The differences in basic concept are extreme and without some background in dome design most engineers fail to look at dome design without introducing built in prejudices. An examination of some basic principles should help in understanding why domes do what they do.

Shape Makes a Difference

Loads applied to slabs and beams create tensile, compressive or bending stresses, which control design. Concrete is an excellent material when used in compression. It has virtually no tensile strength. Steel is an excellent material in tension. A rectangular building utilizing either steel or concrete will result in compromise where concrete needs to be reinforced with steel in tensile areas and steel must be protected from buckling in high compression areas.

To get a sense of what a dome does one should consider what happens when arches are introduced into a design. A properly designed arch will translate external loads into compressive stresses running longitudinally down the arch. Extending this concept to a dome allows one to consider that a dome is similar to an arch that curves in both directions. This understanding can be expanded to visualize the fact that a dome translates external loads into compressive stresses in ALL directions in the plane of the shell. This is what gives a dome the unique capability of carrying stresses around openings that are either planned or the result of damage. This physical characteristic allows the designer to reduce his concern for bending stresses. He needs only to be sure that the design eliminates bending from occurring in the structure.

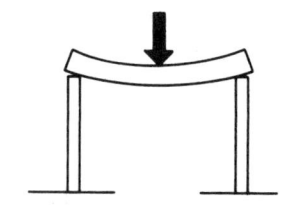
LOADING A BEAM

LOADING AN ARCH

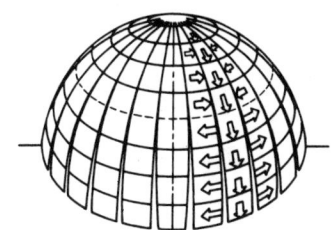
LOADING A DOME

Membrane Theory of Shells

The formula below are based on ELEMENTARY STATICS OF SHELLS by Alf Pfluger, F.W. Dodge Corp. NY 1961.

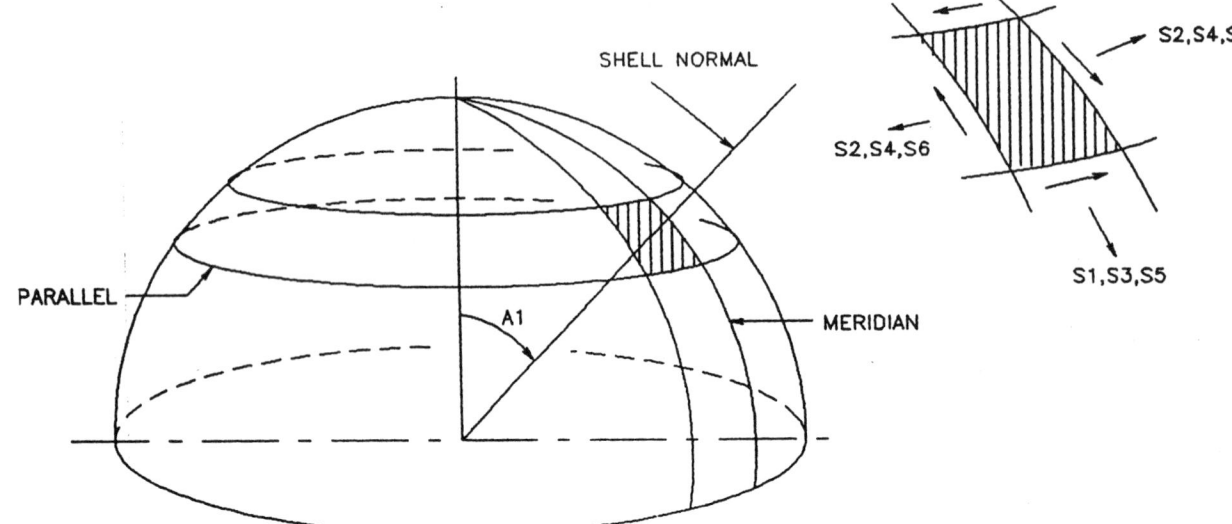

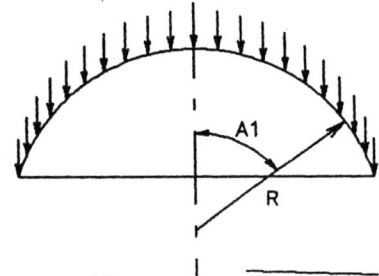

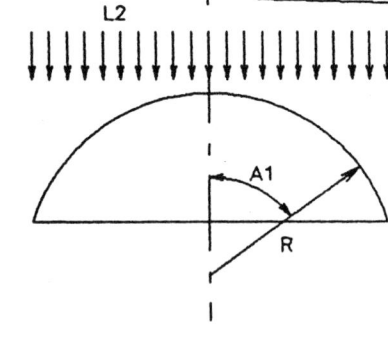

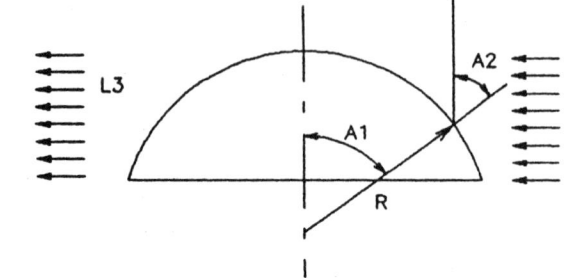

DEAD LOAD CONFIGURATION

S1 = MERIDINAL STRESS/DEAD LOAD

$$S1 = \frac{L \times R}{1 + \cos A1}$$

S2 = HOOP STRESS/DEAD LOAD

$$S2 = L1 \times R \left(\frac{1}{1 + \cos A1} - \cos A1 \right)$$

SNOW LOAD CONFIGURATION

S3 = MERIDINAL STRESS/SNOW LOAD

$$S3 = \frac{-L2 \times R}{2}$$

S4 = HOOP STRESS/SNOW LOAD

$$S4 = \frac{-L2 \times R \times \cos(2 \times A1)}{2}$$

WIND LOAD CONFIGURATION

S5 = MERIDINAL STRESS/WIND LOAD

$$S5 = \frac{-L3 \times R \times \cos(A1) \times \cos(A2)}{3 \times (\sin A1)^3} (2 - 3 \times \cos A1 + (\cos A1)^3)$$

S6 = HOOP STRESS/WIND LOAD

$$S6 = \frac{L3 \times R \times \cos A2}{3 \times (\sin A1)^3} (2 \times (\cos A1)^2 - 2 \times (\cos A1)^4)$$

This does not mean that bending cannot be visited upon a dome design. The size of the external loading can overwhelm the shell's capability to distribute the load and bending (or even puncture failures) can be experienced. Normal roof loads for dead load, snow and wind are usually far too small to have any such effect.

Large equipment loads can induce bending and this must be provided for. Large uniform loads such as experienced on buried structures can also induce bending and the thickness of the shell must be increased substantially to compensate.

The main factor to be considered in the design of a dome is the r/t value.

    r = radius of the dome   t = thickness of the shell.

This value should not exceed 500 for a normal roof design.

Finite element analysis of domes indicates that most of the dome surface contains only compressive stresses, except around openings and where the dome meets its supporting structure or the ground. At these junctures steps must be taken to deal with tensile stresses.

A Tension Ring or Buttresses

The base edge of the dome must be integrated with a tension ring that is capable of dealing with the tensile forces imposed upon it by the dome. In most cases a modest amount of rebar will suffice for a small (up to 100' diameter) dome. When one gets into stadium size domes the tension ring can become a rather massive structure by itself.

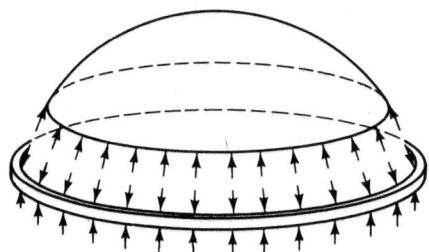

DOME WITH TENSION RING

When dealing with residential structures with many openings being required at the base, the designer can consider using buttresses rather than attempting to complete a tension ring around the entire structure.

If we wish to concentrate on adobe (Superadobe) type structures, the only way to deal with the stresses at the base is with the use of buttresses to eliminate the need for man made materials. This also becomes a matter for consideration in poorer countries of the world where the cost of man-made materials exceeds the financial capability of a country to build enough structures to benefit an adequate number of it's citizens.

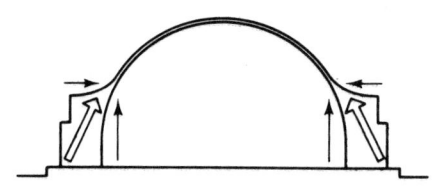

DOME OR VAULT WITH BUTTRESS

Failure

Normal beam and column design has a built in capability for disaster. Every element is designed to cope with the loads being placed upon it by the structure loading or by an adjacent building element. (i.e. column loads are imposed by beams or slabs) Under no circumstance do we design an element to resist the loads which may be imposed by the failure of another element. If a beam fails, the load on that beam goes directly to the adjacent parts of the structure and in almost all cases, this will cause the adjacent structure to also fail. This can proceed into a cascading effect which brings down the entire structure with the failure of one element.

A dome on the other hand will handle a failure in a very different manner. Any point load which induces bending or punching shear which exceeds the shell's capability will cause a hole to be created around the load and the load will then proceed to the ground. The remaining shell will then translate any remaining stresses around the hole with the vast majority of the structure continuing to stand. Impact loads will be handled in exactly the same way which makes a dome an ideal structure for resistance to explosives.

Seismic loads are calculated mainly as a percentage of combined dead and live loads on a structure. Normal column and beam structures are uniform in size and shape as you go up in the structure. This causes massive problems at the base of a structure during a seismic event. Conversely, as you approach the top of a dome the dead and live loads diminish significantly. This factor significantly reduces the actually loads on the base of a dome during a seismic event. In addition the size of a dome increases as you approach the ground and this gives a dome enormous resistance to damage during a seismic event.

Stress Path

Stresses in a column and beam buildings follow a well defined path to the ground which severely limits such a structure when dealing with overloads during seismic events or overloading. A variation of the weakest link phrase comes into effect. "The structure is only as strong as its weakest element." A rectangular building which has one element fail in most cases will suffer catastrophic damage to the entire structure.

A dome on the other hand will transfer stresses in all directions. The dome will attempt to transfer stresses in all directions until all of the elements are loaded to their ultimate capability. In the end a dome will fail locally but not in a catastrophic cascading total failure.

All of the above should be taken into consideration when one is considering a design for adobe (Superadobe) structures. Much of this must be brought to the attention of building officials who are not familiar with domes.

Stabilized or Unstabilized Materials

Stress calculations on vaults indicate that tensile strength is required in the basic material in order to bond to steel reinforcement in tensile areas of the structure. This means that the vault section of a building requires that only stabilized materials be used.

In dome analysis on the other hand, there are vast areas where no tensile strength is required for any purpose. This would allow for the use of unstabilized materials in many areas of a dome. The most obvious exceptions would be in the area of a tension ring and around openings.

All of the above should be taken into consideration when one is considering a design for adobe (Superadobe) structures. Much of this must be brought to the attention of building officials who are not familiar with domes and resist their use.

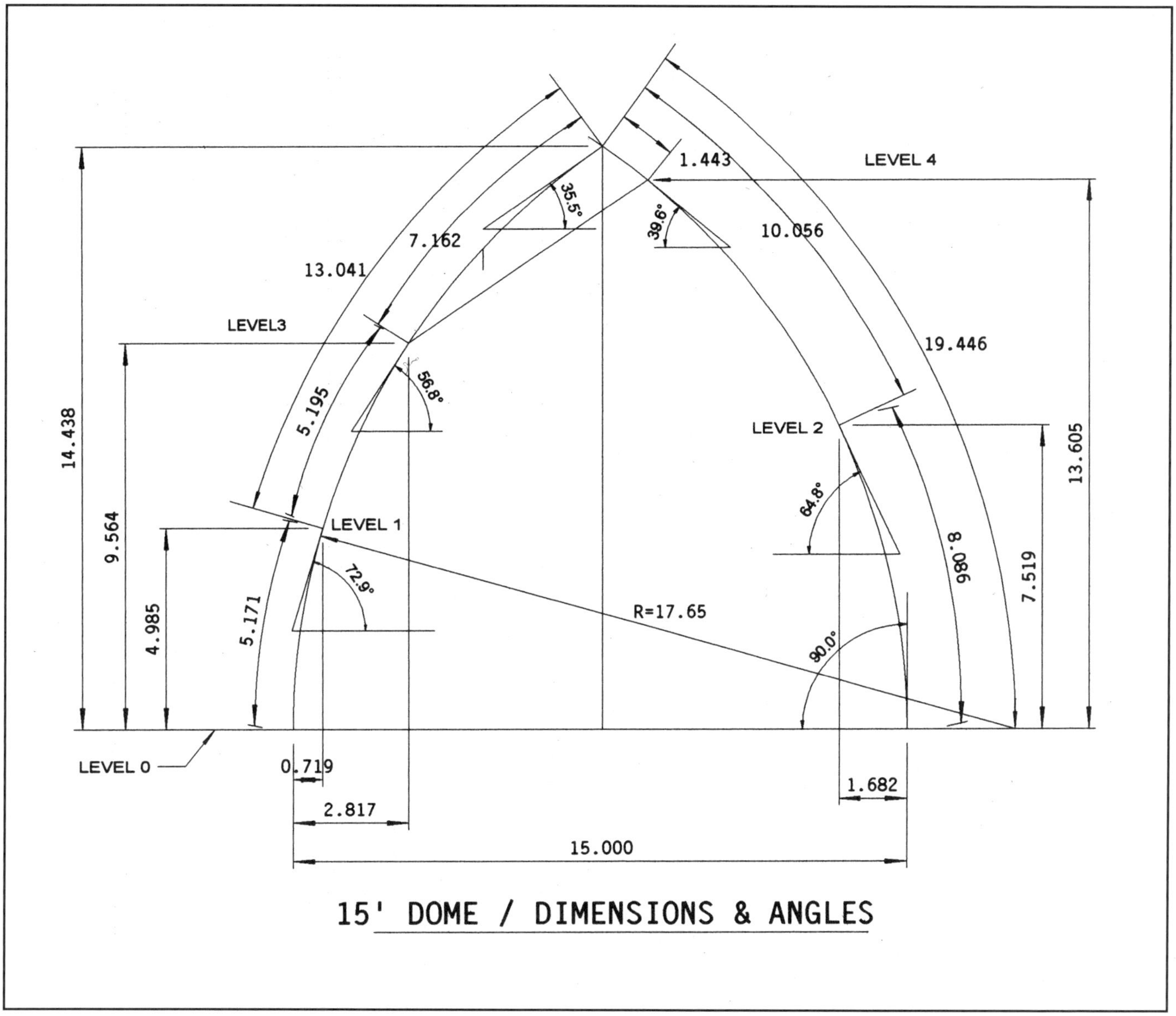

15' DOME / DIMENSIONS & ANGLES

Above and below: Typical diagrams for calculations of a 15 ft. interior diameter dome.

Testing the Domes in Hesperia

In order to obtain a permit for a number of domes to be built for the Hesperia Recreation and Park district, it became necessary to submit full-scale models of the domes to test loading to prove out the accuracy of the design engineering. All the domes were built with unstabilized earth filled bags and barbed wire. No cement (concrete) was used in either the foundation or the dome. These tests were carried out in two phases.

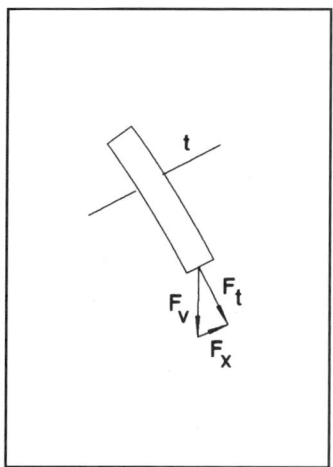

SUPERADOBE - SANDBAG SHELTER

Dead Load, Live Load and Modified Seismic Tests

In order to test an existing dome structure, it was accepted that we would load the surface of the dome with sandbags to simulate 200% of code required live loading. In addition, the loading was place only on one quadrant at a time to simulate the shear stresses, which would be encountered during a seismic event.

To observe if a problem was being developed, a series of deflection observation stations were set up on the interior of the structure. It was concluded that if vertical or horizontal deflections of 1/4" were caused by the loading, the stress on the dome was approaching a serious level. The deflections were recorded at numerous intervals to be able to determine "when" the loading became a problem.

All of the loads were placed on the structures and no deflections of any kind were observed. At a 200% of Live Load level the structures had not even entered a strain area which was observable.

Additional Simulated Seismic Testing

The above tests exceeded the loading required by the code and this loading would have caused the failure of many existing structure in the U.S. In spite of these impressive results the building code officials were still not satisfied that these tests "adequately reflected the stresses which would be encountered in a seismic event".

We then devised additional tests which actually caused shear loads to be applied at the building / ground interface. This testing was accomplished by placing high-tension cables around the base of the structure and tensioning those cables to 150% of seismic loading.

In this test we also set up a series of deflection observation stations on the interior of the structures. It was concluded that if vertical or horizontal deflections of 1/4" were caused by the loading, the stress on the dome was approaching a serious level. The deflections were recorded at numerous intervals to be able to determine "when" the loading became a problem.

All of the loads were placed on the structures and no deflections of any kind were observed. At a 150% of seismic loads the structures had not even entered a strain area which was observable.

It is not difficult to come to the conclusion that for simple structures, it is a great benefit to consider curved or vaulted structures in the design. They will be stronger than conventional design and be able to withstand most of nature's destructive forces.

They are cheaper and can be built with unskilled labor from local materials. Many countries without access to cement, structural steel and the capital to purchase, would be well advised to consider these old but proven methods.

Note: All building permits, blueprints and structural calculations have been updated to the 2007 UBC, 2001 California amendments.

Continuous Inspection & Materials Testing

10826 South Norwalk Blvd., Santa Fe Springs, CA 90670
(310) 941-2990 • (714) 526-8441
FAX (310) 946-0026

December 27, 1995

Mr. Thomas K. Harp
Building Official/Planning Director
City of Hesperia
15776 Main Street
Hesperia, California 92345

Re: Observation and monitoring of load testing for Cal-Earth Domes and Vaulted Structures.

Dear Mr. Harp:

On December 12 and 13, 1995 Mr. Jim Honaker of Southwest Inspection and Testing observed and monitored the load testing of sandbag domes, masonry domes and sandbag vaults structures, located at the Cal-Earth research site in Hesperia, California. The testing procedures incorporated were submitted by P.J. Vittore, Ltd. and were approved by the City of Hesperia Department of Building and Safety and The International Conference of Building Officials (ICBO). Dynamic load testing was performed on the sandbag dome, masonry dome and sandbag vaulted structures, on the center wall and outside wall. Static load testing was performed on the sandbag vaulted structures. Testing was performed by Mr. Phil Vittore of P.J. Vittore, Ltd. and various trained personnel from Cal-Earth.

This letter is to certify that the testing was done in accordance with the procedures submitted by P.J. Vittore, Ltd. and the results were accurately recorded and attached hereto. All tests have exceeded the ICBO and City of Hesperia requirements. Our observation was done as a third party inspection, with no financial or any other interest in the products tested. If you should have any questions, please do not hesitate to call.

Sincerely,

Jim Honaker
Deputy Inspector

Reviewed By,

Steven L. Godbey
President

JH/SLG:jm

CC: Nader Khalili, Cal-Earth

INTERNATIONAL CONFERENCE OF BUILDING OFFICIALS

BUILDING STANDARDS™

SEPTEMBER–OCTOBER 1998
$5.00

Focus on Alternative Building Materials

- Rammed Earth
- Adobe
- Straw Bale
- Sandbag Architecture
- Pozzolans

Earth Architecture and Ceramics
The Sandbag/Superadobe/Superblock Construction System

by Nader Khalili
Cal-Earth Institute
Hesperia, California

and

Phill Vittore
Structural Engineering Consultant
Arlington Heights, Illinois

Nader Khalili, an Iranian-born California architect and author, is the designer and innovator of the Geltaftan Earth-and-Fire System known as "ceramic houses" as well as the Superadobe building technologies. He received his education in Iran, Turkey and the United States, and has been a licensed architect in California since 1970. In 1975, he closed his successful practice in the United States and Iran designing high-rise buildings and journeyed by motorcycle for five years through the Iranian deserts, where he worked closely with local villagers to develop his earth architecture prototypes. His impressions have been collected in his book Racing Alone.

Mr. Khalili serves as a consultant to the United Nations and is a contributor to NASA on construction technologies for the moon and Mars. He is the founder and director of the Cal-Earth Institute, Geltaftan Foundation—dedicated to research and development in earth and space architecture technologies for the moon and Mars.

The views expressed here are those of the authors and do not necessarily reflect the opinion or agreement of the International Conference of Building Officials.

Approximately one third of the people of the world live in houses built with earth, and tens of thousands of towns and villages have been raised practically from the ground they are standing on. Today, world consciousness about the use of natural resources and the new perception of building codes as the steward not only of individuals' safety, but of the planet's equilibrium, are leading us into the new millennium of sustainable living.

In 1984, NASA's first symposium on lunar bases and space activities of the 21st century enthusiastically received the presentations dealing with the utilization of onsite natural resources to construct future lunar and martian habitations. The integration of the ancient technologies of building with earth into planetary construction techniques was presented with the following passage:

> The accumulated human knowledge of the universal elements can be integrated with space-age technology to serve human needs on Earth; its timeless materials and timeless principles can also help achieve humanity's quest beyond this planet. Two such areas of knowledge are in earth architecture and ceramics, which could be the basis for a breakthrough in scales, forms and functions. . . .

The Sandbag/Superadobe/Superblock technology presented here (ceramic structures are also part of generic earth architecture systems) is the spinoff from several consecutive presentations to the space and planetary scientific community since that first symposium in 1984. These concepts have been the subject of an intense research and prototype construction program since 1991. The research and development, engineering, and testing have been carried out at the Cal-Earth Institute under the scrutiny of the Hesperia Building and Safety Department, in consultation with ICBO. Building permits for the stock plans of "Earth One" housing models as well as the Hesperia Lake Museum and Nature Center have been issued, and the Sandbag/Superadobe/Superblock building technology is recognized as a construction system.

A New Approach to Sandbags

Common sandbags and connecting barbed wire, as well as mile-long bags, are referred to here as Sandbag, Superadobe and/or Superblock construction. For centuries, sandbags have been used as elements in building temporary dikes and protective walls in combat zones, as well as in numerous lesser applications. After the structure has served its temporary purpose, the sandbags normally are removed, emptied and discarded.

The Sandbag/Superadobe/Superblock building system builds on three fundamental aspects of historical sandbag modules, resulting in a permanent system of construction:

1. The most serious drawback in the past concerning sandbags as a structural element is that a stack of bags has no tensile capabilities, which has kept structures very low in height. Also, curved, arched or domed structures were impossible without some friction and tensile resistance available.

Superadobe uses four-point barbed wire (or a similar element) between sandbag layers, allowing one to develop the tensile and shear capabilities that have not been previously achievable. The barbed wire element increases the friction factor between the bags and creates tensile resistance in a wall or structural element. It is an important aspect of Superadobe to provide for the

Wall of Superadobe construction—earth-filled sandbags and barbed wire. Coils/courses can be filled by hand or by pump at speeds of 10 to 15 feet per minute, depending on bag width.

transfer of shear stresses from one sandbag to another by using the barbed wire as an interface between the bags, overcoming problems of low shear capability in the earthen fill. The increased capacity of the sandbags, achieved by using barbed wire, creates the capability of designing higher walls and curved surfaces, such as bearing walls, arches, domes and vaults.

2. Previously, sandbags were not considered part of a permanent structure due to the use of loose fill material, usually sand, which can be loaded easily and discarded when the temporary structure is no longer needed.

Superadobe fabric tube or individual sandbags are packed with different mixes of fluent, particulate material. These include earthen, cementitious, organic, manufactured and recycled materials that form into a permanent block.

A model house similar to this one built of Superadobe was constructed and tested as a prototype in Hesperia, California.

3. Historically, the potential deterioration of the bag and the subsequent effect on the structure has precluded permanent structures. Superadobe construction shields the sandbag walls from the elements with protective overlay materials. Additionally, the fill material becomes self-supporting once it has been formed into a block by the tubing. When the fill material is sufficiently resistant by itself, the shielding of durable exteriors is not necessary.

The Sandbag/Superadobe/Superblock system, which has developed out of these fundamental changes during intensive research in the last seven years, is used in conventional structures for foundations (poured within the tubing form), for load-bearing and partition walls in conjunction with conventional roofing systems that bear on a bond beam, also generated by the Superadobe tube itself. Emphasis in this article, however, is given to curved structures, such as domes and vaults, since others are addressing the conventional post-and-beam and rectilinear bearing wall structural systems for earthen construction.

Non-post-and-beam Structural System

Superadobe techniques enable the construction of monolithic structural systems built entirely from earth in curved forms. The sandbag, because of its flexibility, allows the construction of curved surfaces. When using single- and double-curvature compression shells—arch, vault, dome, apse—the majority of conventional roofing systems can be eliminated. In the case of wood construction, this can save up to 95 percent of timber, allowing not only for forest products to be more wisely utilized but also resulting in fire-safe buildings. By working with the principle of gravity, these features can be built without special formwork.

The success of the tested prototypes for California's seismic codes and the resulting permits derive from the following principles:

1. Single- and double-curvature compression shells transfer their stresses along the surface of the structure and not from element to element like column- and beam-type buildings. When a single element in a beam and column construction is overloaded to failure, the loss of that element will create a cascading effect on adjacent elements, causing failure of all elements in the vicin-

Trainee builders work with hand-filled Superadobe. Long sandbags are filled with earth from the site and tamped in place on the wall. Between courses are set two strands of four-point barbed wire.

ity. In many cases, this will cause the entire structure to collapse, as was witnessed in earthquakes in Northridge, California, and Kobe, Japan. Such a structure is only as strong as its weakest element. In a dome, and to a lesser degree a vault, excessive loads on their surface will first cause a puncture failure. This results in the excessive load being shed with only localized damage; the remaining stresses in the vicinity of the failure are transmitted around the failed area, and other loads continue to be held by the structure without any problem.

2. Dead-load and live-load stresses are transferred to the supporting ground, spreading uniformly along the perimeter of a dome or bearing wall. In a beam and column structure, the loads are concentrated and transferred to the ground via a footing under each column. This situation creates the two basic structural problems of differential settlement and frost heaving. These can cause severe localized stresses within the upper structure, resulting in cracking and other failures. For this reason, most foundations are extended to below the frost line to minimize such problems. In a monolithic bearing wall, dome or vault, differential settlement and frost heaving do not pose severe problems. The base of a dome or bearing wall distributes the load of the structure over a much larger area, and local soft spots in the supporting soil will not create a local problem, as local depressions may be easily spanned. The effect of frost can be rendered negligible with correct design when a dome is free to float on the ground.

3. One of the most significant advantages of a domed or vaulted bearing wall structure is its performance in earthquakes. It is difficult to design conventional structures to withstand earthquake stresses. Their basic shape creates a severe problem, as the building weight is either uniformly spread from roof to foundation or, even worse, weights are often larger in the upper floors. With this propensity for overturning, the deeply planted footings and foundations rip apart at the very base of the structure during an earthquake, causing failures rather than preventing them. Modern earthquake design that incorporates foundation isolation does have shifting capabilities, but it is expensive.

Double-curvature compression shell, or dome, of Superadobe construction showing rebar guides for the dome shape (removed later) and a simple plank scaffold. Gravity is the generator of the form.

(continued on page 29)

Sandbag/Superadobe/Superblock: A Code Official Perspective

When architect Nader Khalili first proposed constructing buildings made of earth-filled sandbags, stacked in domes, the building department was skeptical, to say the least. In fact, if we hadn't been trained to be courteous, we would have laughed out loud. How could anyone believe that you could take native desert soil, stuff it into plastic bags and pile them up 15 feet (4572 mm) or more high? Why, if they didn't fall down from their own weight, the first minor earthquake would cause a total collapse, killing everyone inside. How could a responsible building official possibly condone such building code heresy?

Well, Nader Khalili is a very persistent man. Over time, he convinced us that he was going to prove our skepticism wrong, that earth-filled sandbags (now called Superadobe) could be built to meet the rigorous standards of the 1991 *Uniform Building Code*™ (UBC). It all started with Sections 105 and 107, allowing building officials to consider the use of any material or method of construction "... provided any alternate has been approved by the building official" and to require testing to recognized test standards as determined by the building official. Although we had applied these sections numerous times, we had never used them to such an extent on a building so foreign to our codes. To say the least, it was a challenge.

Here's a brief description of what we did. Since we are not licensed engineers, and we don't have one on our staff, we contacted ICBO Plan Review Services to see if they would perform the plan review for our city. ICBO welcomed the challenge, but indicated the same skepticism we shared, since Hesperia, California, is within Seismic Zone 4 and local examples of this type of construction are nonexistent.

After some initial discussions regarding standards, Mr. Khalili submitted plans in November 1992, with the understanding that a testing program would be designed as part of the plan review process. In January 1993, ICBO returned the plans with nine general comments, including a provision to provide a rational analysis pursuant to 1991 UBC Section 2303(b).

At about this time, we were introduced to Mr. Khalili's structural engineer, Phill Vittore. Mr. Vittore, with his partner Morrall Harrington, had designed numerous large thin-shell dome structures in the Midwest and was properly represented by a California licensed structural engineer. Mr. Vittore responded to ICBO's initial comments, and a negotiation began that resulted in the design of a static load test program that was agreed to by this department after our discussions with ICBO staff.

The static load test was designed to add 200 percent of the UBC loading of 20 pounds per square foot (psf) (97 kg/m^2) live and 20 psf (97 kg/m^2) wind load. The first test used an 80 psf (390 kg/m^2) loading of additional sandbags over one third of the exterior surface and, after monitoring, over one half of the exterior surface. During the entire test period, deflection was monitored to verify if ultimate loading was approached. Two domes, one of sandbags and one of unreinforced brick, were tested. Special inspection by a local engineering firm was approved and test results showed "that there was no movement of any surface of either dome structure as a result of the loading described in the test procedure." The domes had passed their first test.

After reviewing the test results, ICBO's Plan Review Services staff felt that the use of the domes should be limited to 15-foot (4572 mm) domes of Group M, Division 1 or Group B, Division 2 occupancies until sufficient monitoring had been completed. Mr. Khalili was principally interested in Group R Occupancies, although he was also proposing the construction of a museum and nature center, a building that would house a Group A Occupancy in a 50-foot (15 240 mm) diameter dome. Because of his desire to build larger structures and house occupancies other than Group M, Division 1 and Group B, Division 2 occupancies, Mr. Khalili notified the city that he would not accept the size and occupancy limitations and would propose new testing to approve the use of larger structures.

After extensive negotiations, which lasted more than a year and included a face-to-face meeting at ICBO headquarters, we agreed to a dynamic test procedure. The procedure involved applied and relaxed loads over a short period of time, with a series of tests with increasing loads until Seismic Zone 4 limits were exceeded. After several months of fine tuning and discussion of "passing grades," the tests and desired results were agreed to. Tests involved three buildings, including the brick dome, the sandbag dome and a sandbag vault structure with 5-foot-high (1524 mm) vertical walls and a barrel vault above. The tests were conducted and monitored by an ICBO-recognized testing laboratory in December 1995, and the required test limits were greatly exceeded. Testing continued beyond agreed limits until testing apparatus began to fail. No deflection or failure was noted, however, on any of the tested buildings.

With these results, the plans went back to ICBO, and after final plan check comments were satisfied, ICBO recommended the plans for approval in February 1996. Our skepticism had long since vanished, as we had seen this style of building meet and exceed the testing of rational analysis as required by our code. Mr. Khalili had succeeded in gaining acceptance by the City of Hesperia for a building made of sandbags filled with earth. It is a testament to Mr. Khalili's perseverance and to the flexibility of the UBC.

—*Tom Harp*
Building Officer/Planning Director
City of Hesperia, California

—*John Regner*
Senior Plans Examiner
City of Hesperia, California

(continued from page 27)

A dome or bearing wall built on a floating foundation, the base isolated by a layer of gravel or sand, provides the ideal earthquake-resistant structure. The continuous or ring foundation can slide across the moving ground, while the upper structure, which diminishes exponentially in mass toward the apex, performs as a unified monolithic piece, eliminating local failure higher up the building.

Structural tests performed during intensive prototype research in Hesperia addressed both live-load and dead-load, static as well as dynamic loading forces. The structural engineering and testing procedures were conducted by Phill Vittore of P.J. Vittore, Ltd. The successful results were documented by an ICBO-approved testing laboratory, Southwest Inspection and Testing.

What follows is the chronology of testing, which includes static and dynamic load tests of the prototype Sandbag/Superadobe/Superblock and masonry structures.

CODE TESTING CHRONOLOGY

Tested models:

1. Two domes constructed with standard sandbags and barbed wire, filled with pure desert sand/earth excavated in situ. The sand/earth used in the first dome was dry, while for the second dome it was dampened with water before filling the bags.

2. A vaulted structure consisting of three adjoining vaults constructed with pure desert sand/earth excavated in situ. The walls were built with standard sandbags and barbed wire, as well as long tubular bags (Superadobe/Superblock) filled with dampened sand/earth. The roof material for the 4-inch (102 mm) slab was the same sand/earth stabilized with 7 to 10 percent portland cement.

3. A 4-inch-thick (102 mm) perforated dome constructed with standard fired clay bricks and cement mortar with no reinforcing.

June 1992. Two earthquakes centered in Big Bear and Landers, California, measuring 6.9 and 7.4 (respectively) on the Richter scale, affected the Hesperia site of the Cal-Earth Institute, with the completed brick masonry dome and the Sandbag/Superadobe/Superblock dome under construction.

October 1992. Plans and engineering for two completed prototype domes in Superadobe and masonry were submitted to, and reviewed by, the Hesperia Building and Safety Department in consultation with ICBO.

September 24–27, 1993. Live-load tests to simulate seismic, snow and wind loads. Static eccentric loading of both domed structures to 200 percent of code requirements. Monitored by independent engineers from the Inland Engineering Corporation. No deflections were observed.

February 18, 1994. *Uniform Building Code*™ (UBC) Group M, Division 1 and Group B, Division 2 structures were allowed in Hesperia as part of a prototype program.

October 27, 1994. Plans and calculations for the Hesperia Museum and Nature Center, to be constructed using earth and ceramic architecture technology (Superblock/Superadobe and masonry/ceramics), were submitted to, and reviewed by, the Hesperia Building and Safety Department in consultation with ICBO.

December 12–13, 1995. Simulated dynamic and static load tests were performed on all structural prototypes relevant to the project to establish their safety for all UBC occupancy categories. Tests were performed on Superblock/Superadobe dome type (double-curvature compression shell), vault type (single-curvature compression shell) and masonry dome type. Testing monitored by Southwest Inspection and Testing.

December 27, 1995. Report by Southwest Inspection and Testing to ICBO and the Hesperia Building and Safety Department concludes that all tests have exceeded ICBO and City of Hesperia requirements.

March 7, 1996. Construction permit was issued by the Hesperia Building and Safety Department for the Hesperia Museum and Nature Center.

January 8, 1998. "Earth Architecture—Environmentally Friendly Housing Types: Superadobe Model House Plan Permits for 3- and 4-bedroom Models" was issued by the Hesperia Building and Safety Department as stock plans for the Earth One house and variations on the plans and designs.

Over this period, climatic stress to the prototype structures was monitored in the harsh high-desert climate zone, including flash floods, heavy driving rain, dry heat up to 115°F (46°C), freezing and snow, and high winds.

Universal Applications

Modern computer software now allows for structural design analysis on an individual basis. The computer will also permit the utilization of the Sandbag/Superadobe/Superblock systems in space and planetary construction based on performance programs, such as finite element analysis. The construction of infrastructures, structures and shielding elements, such as for thermal, radiation and/or impact shielding on the moon and Mars, would otherwise imply costly transportation of building materials into outer space. The utilization of *in-situ*, minimally processed materials, is crucial to space exploration.

Flood control; erosion control; stabilization of waters' edges, hillside slopes and embankments; and retaining walls, landscapes, and infrastructures are applications in which the Sandbag/Superadobe/Superblock system has shown great potential.

Individuals are enabled once again to build their own homes without the use of heavy equipment, with materials native to the country of use. All the skills required are simple and can be acquired by anyone who wishes to learn them. The Sandbag/Superadobe/Superblock system can use existing contractors' machinery, such as concrete and gunnite pumps, to mechanize the packing of the fill material into the bag forms.

Sandbag/Superadobe/Superblock has been used internationally by the United Nations for emergency housing prototypes and is currently in limited use on several continents and under construction in several states in the United States. Further permits are being sought with other building and safety jurisdictions in California and the Southwest. If integrated into the 2000 *International Building Code*™, it will serve internationally for many building types, including that of emergency housing construction, and be used extensively within the United States for standard housing. ■

Reprinted from the September - October 1998 issue of Building Standards, Copyright 1998, with permission of the International Conference of Building Officials.

Above: Structure of the coiled Superadobe Eco-Dome.

*"an arch is like a prayer
its strength is in its unity
its beauty in its repetition"*

A·P·P·E·N·D·I·X II

ARCHES
A TIMELESS PRINCIPLE

Above: Earth and arch.

Right: Water and vault.

Arches in Nature

To understand the vocabulary of the timeless elements of this architecture, we need first to learn about the Arch. There is no straight line in the universe, all is ultimately curved, as Einstein showed. The architecture which I have been doing for the last several decades is called "Archemy" which is a fusion of alchemy and architecture. When the universal elements of earth, water, air, and fire generate a form or structure it is curved in nature.

The rainbow in the sky, that we teach children at Cal-Earth to imagine as their house, is nothing but an arch of sunlight through the lense of air and water.

An arch left over from millenia of erosion, such as those in "Arches National Park" in Utah, is the inherent structure in the earth. It is created by gravity, friction and the "angle of repose" of the material. Over millennia, the ocean carves the shoreline and creates arches made of rock. Drop by drop water erodes mountains leaving behind the arched and domed spaces we call caves. These arches and domes carry the load of the entire mountain above.

Volcanos create Lava Tubes, long arched tunnels, as they spew out rivers of molten earth. In the cool atmosphere the lava forms a crust, while the hot, fluid interior continues to flow, molded by gravity and the material's inherent structure. When the inner flow is exhausted, all the lava has become a self-supporting arched tunnel, as you can see in the "Craters of the Moon" National Park in Idaho, and also in northern California. Similarly, a vault is a series of arches placed back to back to create a tunnel. The most ancient leaning arches which I saw in the archaeological site of Choghazanbil in southern Iran, are over 3,300 years old and create vaults.

The arch is the ideal form for earth architecture, and as Architect Lois Kahn once said,

"Ask bricks what they want to do and they'll say, 'Make us into an arch'. Try to sell them a lintel and they'll say, 'Make us an arch'."

Below: Water and arch.

Left: Volcanos or eroded mountains are all dome shaped structures.

Principles of Arches

In nature, tension and compression mirror each other, but they both make arches. A spider's web, a hanging chain, a cable bridge like the Golden Gate bridge in San Francisco are structures in tension, whilst a lava tube or magma vault, a masonry arch, or a Superadobe arch are structures in compression.

Our human spine also creates an arch. The many vertebrae are its building blocks. To experience how an arch feels in gravity, when children come to Cal-Earth we ask them to make a human arch. Immediately they have understood the idea, they play games making human domes and vaults, sending one child through the long tunnel made by many human arches. The vertebrae and muscles make an arch when we bend down, balancing tension and compression.

An arch is a combination of vertical and horizontal forces. Thus water can make an arch that we call a fountain, and a wave can make a dynamic tunnel.

The ingenious Superadobe, sandbag and barbed wire system has tension and compression elements together. The earth works in compression and the barbed wire works in tension.

Above: The vertebrae of the human spine will curve into an arch.

Left: Children make a human arch.

SUPERADOBE - SANDBAG SHELTER

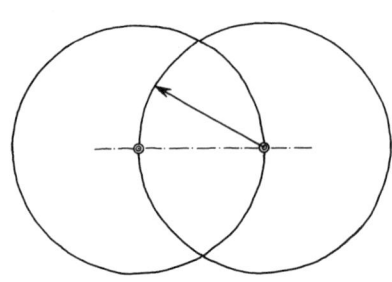

Above: Ogival or Lancet arch as the intersection of two circles.

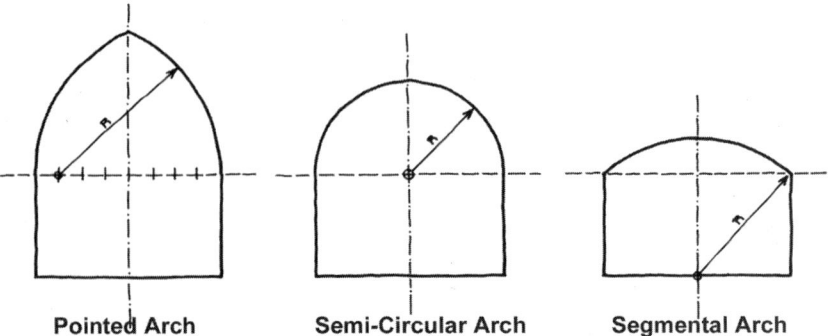

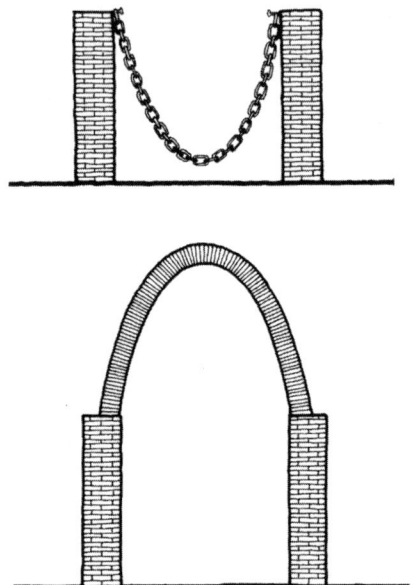

Above: A hanging chain in tension is reversed to become a catenary arch.

Arch Geometry

What are the type of arches that are most used? The semicircular arch is one that everybody knows, with one radius from the center. A segmental arch is a flatter segment of this. The catenary arch is the inverse of a hanging chain where total tension in the chain becomes total compression in the catenary arch; it is the strongest and most stable arch in gravity. Pointed arches are usually tall and are made by inersecting two circles. A beautiful example is the eight-section arch, traditional in the Middle East and Gothic architecture. The Ogival or more pointed Lancet arch is one of the most suitable for Superadobe.

The conical, pointed arches, and catenary arches are closest to the earth's "angle of repose", which is the natural cone formed when lava makes a volcanic mountain.

By repeating an arch back to back we make a vault, by rotating it we make a dome, and by raising it from horizontal to vertical we make an apse (half a dome). With these forms we can build a home, a community, even an entire town. Therefore, when we learn how to build an arch we have learned the entire alphabet of this architecture.

Right: Arches can be built with short or long bags.

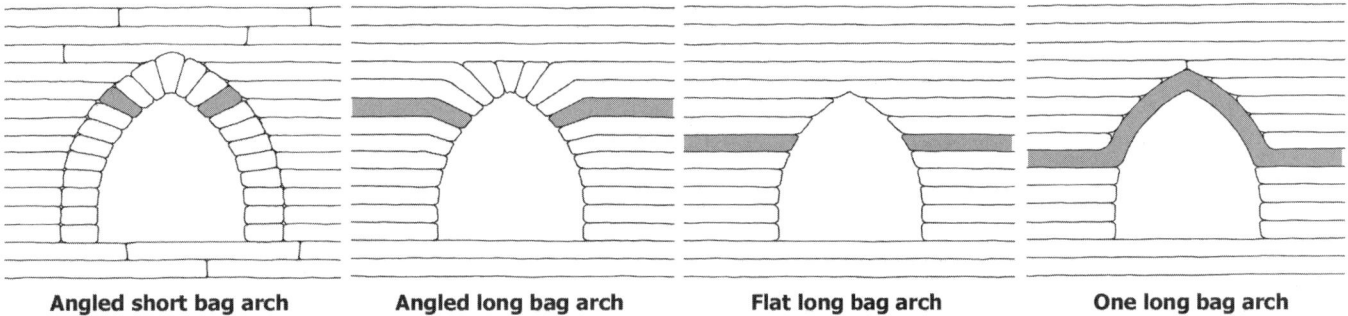

| Angled short bag arch | Angled long bag arch | Flat long bag arch | One long bag arch |

Superadobe Arches

The best way to make any opening in a Superadobe wall is with an arch. Arches built with bags create door and window openings for emergency shelters, and other larger and more permanent buildings. The flexibility of the Superadobe system allows the builder to form an arch in four or more ways. Yet an arch is still an arch, whether it is built with long bags or short bags, laid flat or at an angle.

There are four basic techniques used to build an arch which are:
1)   angled short bags (like a traditional brick arch)
2)   angled long bags
3)   flat long bags (corbelled)
4)   arched long bags

Deciding which arch forming technique to use depends on the skill of the builder and the availability materials to make forms. Choosing the right shaped arch depends on the size of the opening and the use of stabilized earth mixtures. The size of the arch is determined by the design of the building.

Above: Arch with flat long bags.

Left: Arch with angled short bags.

Right: An arch with four bags of buttressing on both sides.

Above and right: Building up both sides together on the form. Note the barbed wire ring on the right, placed between the bags.

Building Arches With Small Sandbags

The best way to learn the spirit of arches is to start practising hands-on, learning with all the senses working together. And the best lesson starts with a "dry-pack arch", so called because it uses only the dry materials of small sandbags filled with dry earth and small pieces of barbed wire. Thus you will learn by experiencing the relationship between the shape of the arch, gravity, and friction.

To build an arch with sandbags you will need a form. It could be of arched plywood or sheet metal, which you will raise several inches above the ground on wedges (wood, or other removable pieces). Also imagine you are in an emergency or in the desert, on the moon or Mars, with no plywood form. Then you will make the form with the same earth-filled bags.

To start, fill the small bags with earth, fold and compact them

Left: Filling and compacting the top bags at the same time

against the form. After compacting/tamping each bag with a brick, place on it a piece of barbed wire or other prickly element, which will increase the friction between it and the next bag.

You must build up both sides at the same time, as the children's "human arch" demonstrates. If one child were gone from the arch the other child would fall over!

As you compact more bags on top of each other into a curve you may see that the bags start to bend away from the lower part of the form. This is because the increasing weight above is creating an outward thrust, due to the shape of the arch. A tall, pointed or catenary arch creates less thrust than a semi-circular or flatter segmental arch. The outward thrust creates tension which must be resisted by the weight of a buttress, composed of more filled bags.

Fill and compact the buttress bags up against the arch, as far up as needed for the shape of your arch.

Above: Tamping the bags against the form using a brick.

Left: Making an arch with no plywood form, using a form of filled sandbags.

SUPERADOBE - SANDBAG SHELTER

Right: Removing the bags which made the form.

Above: Testing the dry-pack arch by standing on it.

Right: Testing a cement stabilized earth arch by loading with sandbags. The arch itself is made of a long, single bag.

When building arches, domes and vaults with Superadobe, the ground itself may be the best buttress, if the structure is sunk down a foot or more into the ground. When built above ground the buttress bags can be designed into seating and landscaping elements attached to the base of the structure.

After building the buttresses and most of the dry-pack arch, you will need to fill in the last bags to close the top. Use the Superadobe flexibility to fill and compact the last three or four bags together. The piled up bags do not become an arch until the last bag is wedged into place, hence the expression "the keystone" used for masonry arches. The key bag is "key" to achieving an arch.

When the arch is complete remove the form; empty the sand from the bag form, or remove the wedges from beneath the wooden form.

Left: Multiple arches being tested with symmetrical and asymmetrical human loads.

Testing the Arches

Firstly, you can see if your arch stands by itself. If it is strong, then test it by loading with more sandbags. As you can see some students prefer to stand on their arch to perform a basic live load test, and experience the limits of their arch of dry sand, bags, and barbed wire. This valuable practice brings you closest to experiencing the force of gravity and loading on your structure. Symmetrical loading is when all the people stand symmetrically. Eccentric loading is when everybody moves to stand on one side only. This is more difficult for a dry-pack arch to resist, and it may start to move. Dynamic loading means your arch is pulled and pushed from the side, as in an earthquake.

The first sandbag arch at Cal-Earth was built in June of 1992 and is still standing after fifteen years. Structural tests on Cal-Earth's prototype domes and vaults consisted of both static and dynamic loading. Some of the first tested buildings were also built with dry/damp desert earth, bags and barbed wire.

Above: This heavily loaded dry-pack arch has a steel cable like a belt from one end of the arch to the other. The cable is used for post-tensioning instead of buttresses.

Left: Testing a dry-pack long bag arch.

SUPERADOBE - SANDBAG SHELTER

Right: Building a brick dry-pack arch over a form raised up on bricks wedges.

Brick Arches

Dry-pack brick arches are also valuable to practice. Place the bricks over a wooden form which is raised up on wedges. Make sure the bricks touch at the inner edge, with no sand in between. Place dry sand or small pebbles between the outer edges. If the bricks don't fit perfectly when you reach the top, share the gap equally between the top three or four bricks and fill with sand and small pebbles.

Try making arches with different shapes like a catenary arch, a semi-circular arch, and a pointed arch. Even a flat arch is possible if it has enough buttressing. I have seen large domes and vaults in the deserts of Iran built with bricks entirely into the ground, which is its buttress. They achieved 20 ft. spans and more with a very shallow rise, almost a flat arch.

Above: Building up both sides together.

Right: Placing the last brick (the keystone) into the arch.

Left: The form is dropped by removing the wedges.

Gravity and Arches

Gravity is a constant force which pulls all structures down to the ground. A flat beam or roof is always fighting against it, but sooner or later gravity will win. By curving a flat beam into an arch we change tension into compression, and therefore an arch, vault or dome becomes strong because it works in harmony with gravity. In your sandbag or brick arch gravity pulls on every element and packs them tightly together, making the arch stronger.

Therefore, the strength of arches, vaults, and domes co-exists with their symmetry in loading and being in harmony with gravity.

Left and above: Completed arches for smaller and larger sizes.

SUPERADOBE - SANDBAG SHELTER

Superadobe long bags create opportunities for sculpting every opening into an elegant arch which can be left exposed.

"unity within multiplicity is achieved by structural form and the integration of compression (sandbags) and tension (barbed wire)"

A·P·P·E·N·D·I·X III

SMALL BAGS
EARLY WORKS

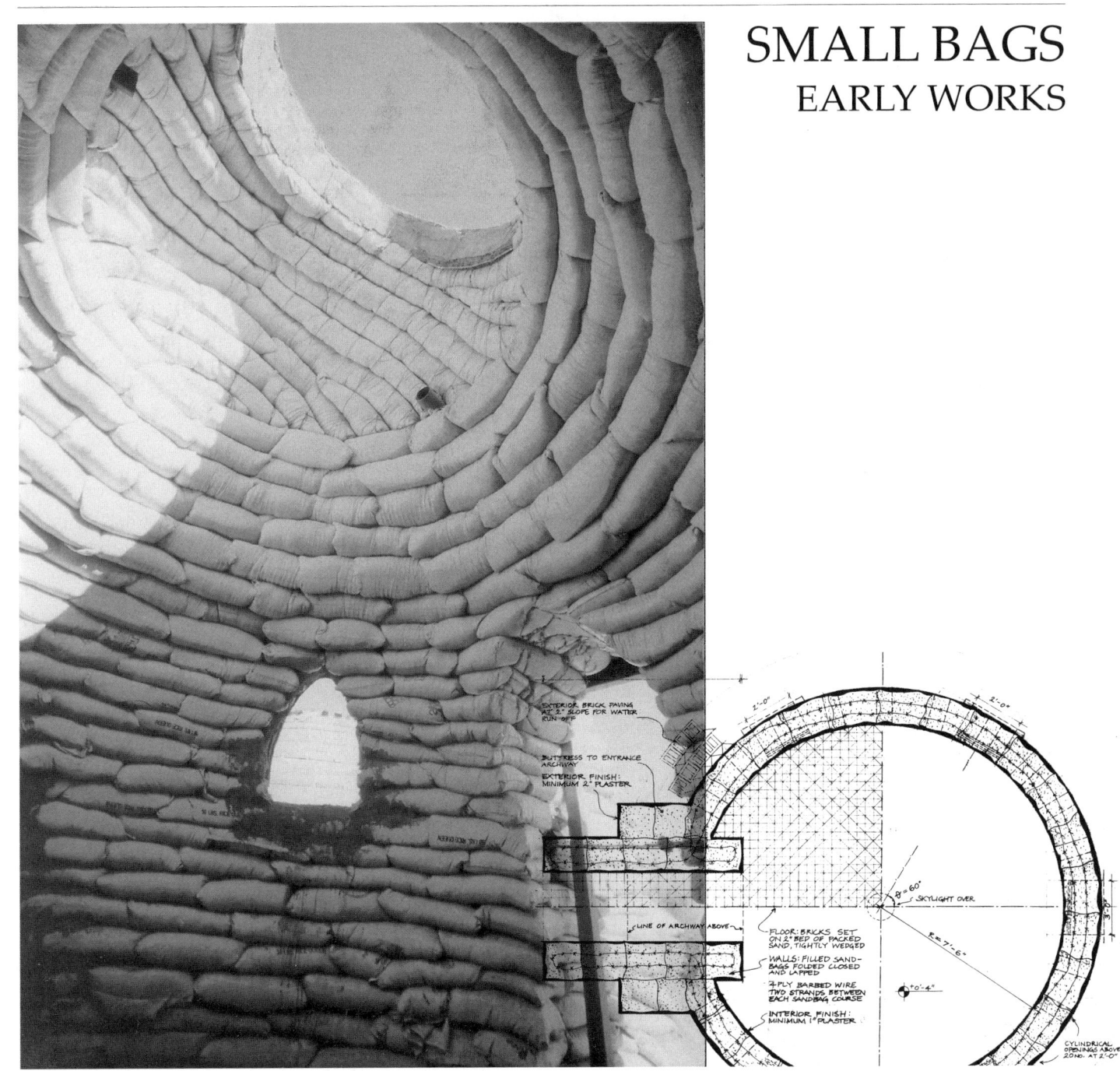

SUPERADOBE - SANDBAG SHELTER

Above: Women and children learn how to build an emergency shelter with standard sandbags.

Standard Sandbags

Early research at Cal-Earth Institute was carried out with small standard sandbags. The goal was to build safe and simple models and methods which can be followed by anyone to build a shelter using the earth under their feet.

In 1993, an earthquake in Gujarat, India killed thousands of people. Cal-Earth's first sandbag dome prototype had been completed in 1992, but the construction technique was not yet practical. The heavy sandbags had been lifted into place, and it was risky trying to angle the bags like blocks. Unable to help, the Gujarat earthquake was the catalyst for an important leap in research; laying the bags flat and stepping inwards with every row, called corbelling.

This technique was tested out at that time on the simplest temporary shelter, which was built using standard sandbags, filled with earth and placed in rings to form a cone. It was small enough not to need barbed wire. This shelter took one person one day to build during the apprenticeship program.

Its small size allowed one or two people to lie down inside. The interior could be warmed in winter by body heat and candles. It required about 200 sandbags and desert soil, and lasted three years in the California desert climate before returning naturally back to earth.

Step 1: A ring of sandbags is made, big enough for a human to lie down inside. A stick and a rope is used to draw a circle. The door was optional, entering from the side or from the top.

Step 2: The bags are stacked in smaller and smaller concentric circles. The ends of the filled bag are folded under so that the dry earth does not spill out. Human feet compact the earth.

Step 3: A conical dome is the final shape. The shelter was warmed by a few candles and the human body.

SUPERADOBE - SANDBAG SHELTER

The first completed sandbag dome prototype.

Domes and Vaults with Small Bags

For larger structures, the small sandbags were combined with four-point barbed wire to create domes and vaults. For small temporary structures up to 10 ft. diameter it was possible to build a dome without the barbed wire.

By filling the bag in place on the wall in rows, lifting heavy bags was eliminated so that women and children could also build. Each row formed a stable platform on which the builder could stand or sit while building, and thus the need for scaffolding was eliminated.

Therefore, buildings with small bags are constructed in the same way as those with long bags. The same rules of structure for domes and vaults apply for short or long bags, for foundations, walls, windows and doors, roofing and finishing.

The larger early prototypes, were also built with unstabilized desert earth - up to 14 ft. diameter for the sandbag domes, and 12 ft. span for the vaults (the walls were built with unstabilized earth bags and barbed wire, whilst the vaulted roofs were built with 4 inch thick stabilized earth slabs, anchored to the walls with 2 ft. long steel reinforcing bars). A 20 ft. diameter brick dome prototype, called "Rumi dome" was also built.

Details of dome construction with earth filled small bags (no stabilizer).

Above: Placing the barbed wire.

Right and below: Constructing by standing or sitting on the rows of bags.

SUPERADOBE - SANDBAG SHELTER

Above: Plastering the first sandbag dome.
Below: Arched window of small bags.

Tested Structures

Three aspects of the Superadobe system increase their safety; building an accurate curvature, stabilizing the earth mixture, adding the 4-point barbed wire. By choosing unstabilized desert earth/sand for these larger prototype structures, and by omitting the barbed wire in some of the smaller domes, it was possible to observe the limit of the system.

The early prototype structures were tested at full scale to verify their safety in earthquakes, especially for California's highest seismic zone 4. Plans and details were submitted to the city building department, along with engineering calculations for the expected stresses during a seismic event. After years of dialogue and a successful series of tests, Hesperia Building and Safety department, in consultation with ICBO (International Conference of Building Officials), issued permits for construction of a museum and housing.

From that time onwards building permits, impossible for earth architecture in California for over 50 years, became a reality. Yet every permit took time and was gained with extra work to prove the safety of these structures time and again. As more buildings are built and are recorded by local authorities, additional data must establish this technology among the most accepted methods of construction.

Above: Inside the first sandbag dome; from left, Eric Lloyd Wright, Nader Khalili, Mayor Mike Lampignano.

Above: Tested sandbag dome exterior.

Below: Early sandbag dome under snow.

SUPERADOBE - SANDBAG SHELTER 273

Right and below: Demolition of a structure was very difficult, even by bulldozer.

Test Conditions

The harsh climate of the Mojave desert in California, and the nearby San Andreas fault, provided the following real life test conditions for the structures year after year:
a) Desert thunderstorms with flash floods
b) High summer temperatures (over 100 degrees farenheit)
c) Below freezing winter nights and snow
d) High speed winds and sand storms
e) Two major earthquakes (1992 and 1999)

Some of the early prototypes had no foundations except for the unstabilized earth bags, and were not waterproofed under the foundations or on the exterior. These were tested by rain and some settlement, proving the general stability of the forms of domes and vaults, and the importance of the barbed wire.

Generally, the bags were plastered over and stayed intact, but in some test structures they were left to degrade in the sunlight, leaving behind only earthen material, barbed wire, and fabric pieces.

All the domes and vaults were observed and documented over a period of 5 years, and most are still under observation at this time, fifteen years after construction.

In evaluating their perfomance we noted the following:
1) Being symmetrical was essential for domes and vaults with low or no stabilization. Asymmetrical unstabilized earth structures can become unsafe if the structure becomes saturated with rainwater.
2) The barbed wire helped to compensate for some asymmetry.
3) The barbed wire, if placed towards the center of the bag and not exposed to the outside, did not oxidise/rust and remained shiny as new despite being sometimes wet from rain.

Left: Flooded sandbag dome. Only the plaster melted down, but the sandbags remained intact.

Below: Laying foundation bags.

4) Buttressing was essential for large openings such as doors, and for the entire structure when larger than 10 ft. diameter.
5) The small bags, when combined with barbed wire, structural symmetry, and buttressing were as effective as long bags.
6) Unstabilized earth structures last as long as they are plastered and waterproofed. Most standard sandbags are rated at 300 UV (ultra-violet) hours. Longer lasting bags are available at higher cost.
7) Demolition of a 15 ft. dome was very difficult, even by a bulldozer because of the geometry of the building and the barbed wires, which would not allow the bags to seperate. The barbed wires were cut before a row could be pulled down.

Left: After 300 UV (ultra-violet) hours, the bags begin to disintegrate. After three and a half years entirely exposed to the elements this temporary structure had reached this stage. It would have lasted with the addition of plaster, as has been demonstrated by Cal-Earth's structures since 1992.

Above and below: The rectangular rooms of the 3-Vault prototype built with small bags.

Availability and Types of Bags

In cases where long Superadobe bags are not available, safe structures can still be built with standard small sandbags. They may not be as fast as long bags, and are less suitable for mechanized filling by pumps, but they are often easier to find, and easy for one person to build.

A variety of small bags exist in the marketplace, which are used by different industries for packaging. Some of the most commonly sold small bags are woven polypropylene, jute "gunny sacks", and cotton sacks. A good place to start looking for bags is the local fire department, which will use sandbags for flood control. The bag must be strong enough to contain the tamped earth, and allow the barbed wire to grip. It must also allow the earth mixtures to cure and to dry out. Of course one must always remember that the long bags are nothing but the sandbag tubes before they are cut and sewn into small sizes. Therefore, if small sandbags can be found there must also be a source of long bags.

Above and below: Completed 3-Vault house with two wind-scoops and reptile finish, after twelve years. Right: Plan of the 3-Vault house.

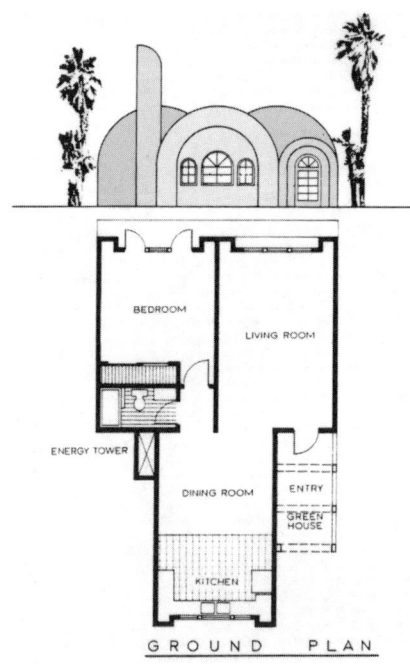

SUPERADOBE - SANDBAG SHELTER

The "Mars One" dome (top) and the 3-Vaulted House house (above) and both built with small bags and barbed wire.

*"the point of unity
generating timeless spaces"*

A·P·P·E·N·D·I·X IV

COMPASS

This chapter shows you how to make and use a compass to build a dome. We recommend that you study this chapter to the end before starting to make one.

A COMPASS is a tool to draw a two dimensional circle on paper or on the ground.

A DOME COMPASS is a tool which works the same way but draws a three dimensional hemisphere (half a sphere) like a bowl; the height is the same as the radius.

The hemispherical shape is suitable for domes made of bricks, blocks and adobes. This compass can be a simple piece of chain or pipe, whose length does not change, and which turns around a point at the dome center.

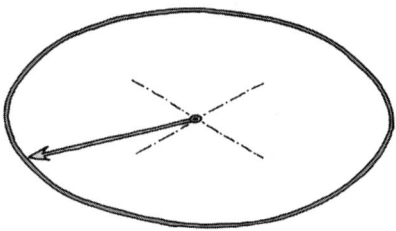

COMPASS

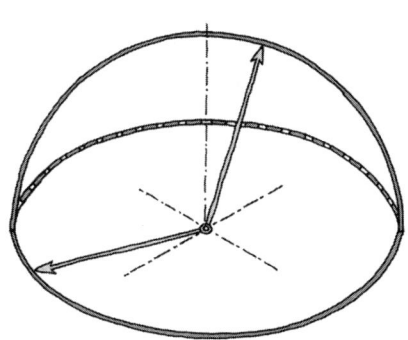

DOME COMPASS

Left: Hemispherical Brick Dome, at the Hesperia Museum.

SUPERADOBE-SANDBAG SHELTER

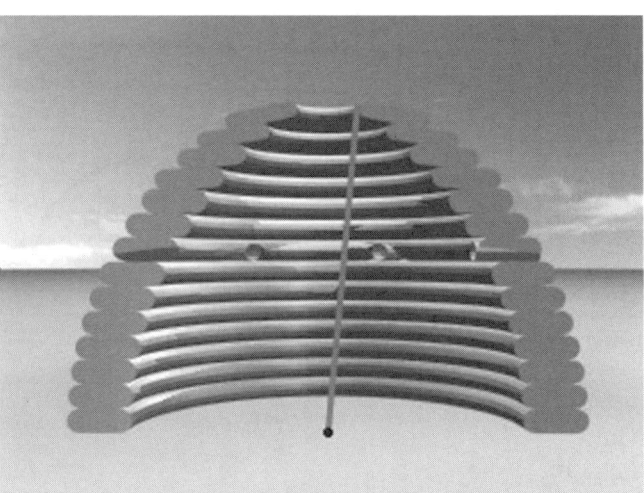

Above and below: The Superadobe Lancet Dome Shape.

A Superadobe dome is not a hemisphere, but it has a taller "Lancet" or "Ogival" shape because of the structure and method of building. The compass for a Superadobe Dome is called a Lancet Arch compass.

To draw the shape of a Lancet Arch on paper or on the ground:

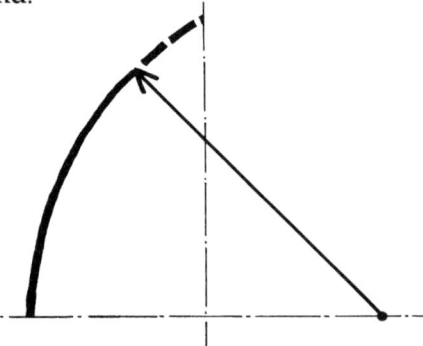

First, we fix a compass on one side and draw one half an arch.

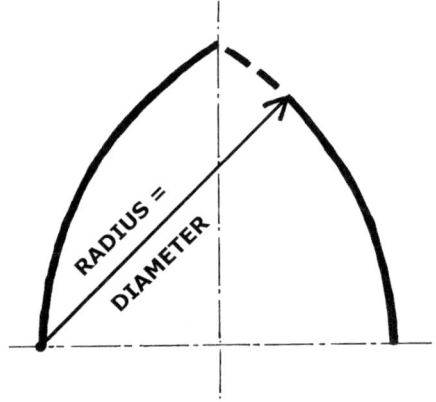

Then we swap the compass to the other side and draw the other half of the arch.

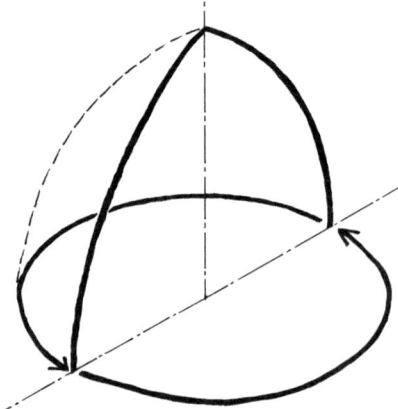

When we rotate the arch around its center we make a Lancet Dome.

A LANCET DOME COMPASS can be made to build a Superadobe Dome with two pieces of chain.

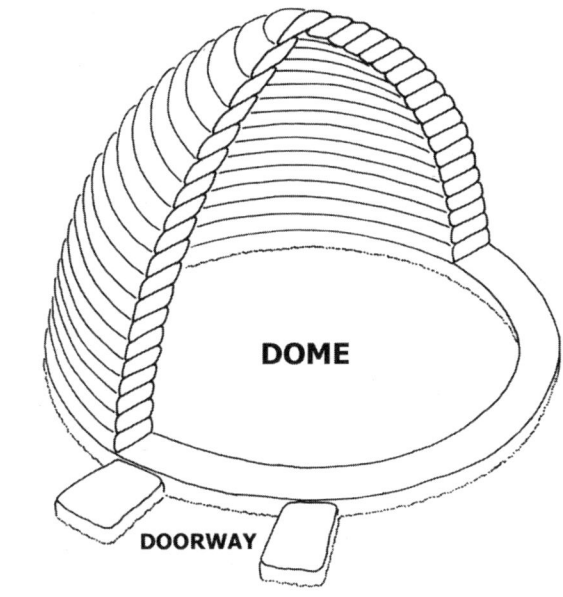

1.

A section view of a Lancet/Ogival Dome made from rows of Superadobe coils.

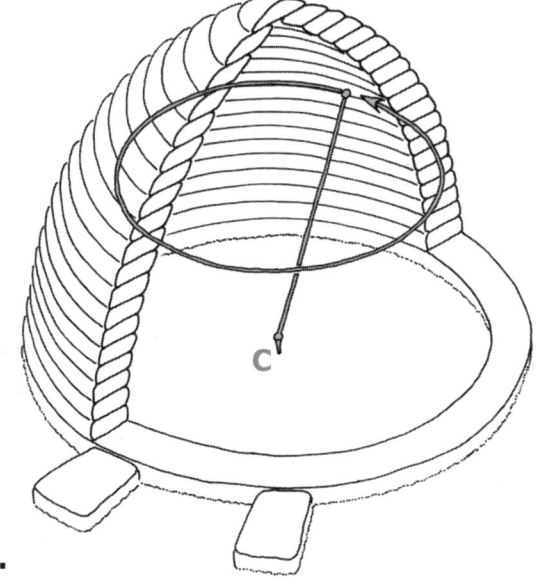

2.

The Center Compass (C) is the chain that rotates around the center and makes each Superadobe row. It gets gradually longer to construct higher rows. To know how long to make the center compass we use the Height Compass.

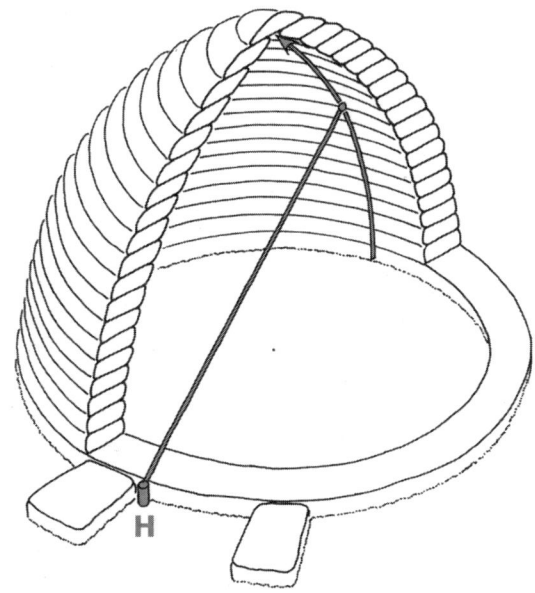

3.

The Height Compass (H) draws half an arch. It is the chain for controlling the shape of a lancet/ogival arch. It always stays the same length, which is equal to the dome diameter at the wall center. For practical reasons we fix the compass on outside and draw the arch to the inside face. The height of the finished dome is approximately equal to its diameter.

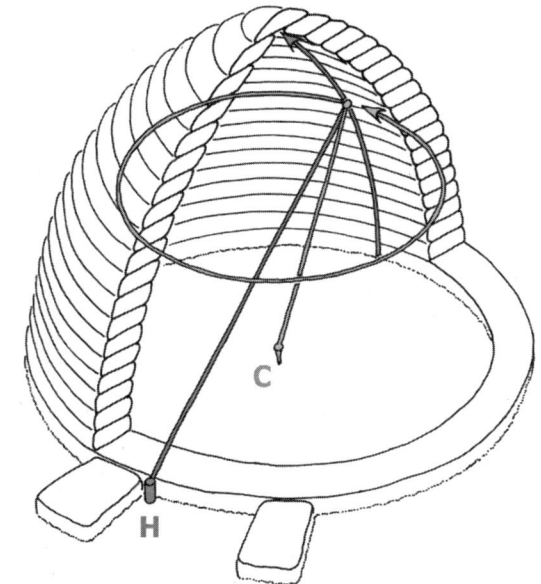

4.

As we begin constructing any row of Superadobe we bring the two chains together and adjust the Center Compass length by matching it with the Height Compass. Therefore, at every row, the Center Compass increases in length to match the Height Compass. Once the Center Compass length is set, the Height Compass is not needed any more for that row.

SUPERADOBE-SANDBAG SHELTER

Above: Step 1. A post with a turning ring on top is being hammered into the ground. (The center post could be a pipe, metal stake, wooden post or other material).

Below: Step 2. The Center Compass chain is attached to the turning ring on top. (The Center Compass could be a long pipe, a cable, or chain).

Right: Step 3. The chain is pulled tight and a circle is drawn on the ground using a rod or stick. Lime or chalk powder is used to mark the circle clearly.

How to Set Up and Use a Two Chain Compass

Step 1) A dome starts with a center. Set a post into the ground for the center of your dome. This center must be firmly in place and should not move until the dome is finished.

Step 2) Attach a chain to the center, which we shall call the Center Compass. Use a lightweight chain that doesn't wrinkle, and attach with metal rings, nails or any other suitable method.

Step 3) Pull the chain tightly across the ground to measure the inside radius distance of the dome along the chain and mark the point with a steel ring or similar. Push a small rod (a piece of steel bar or a screwdriver) through the ring and, while pulling the compass chain tight, draw a circle on the ground. Now you are using a chain, ring, and rod as the Center Compass.

Step 4) Draw another larger circle representing the width of the Super-adobe bags which will make the wall thickness. For example, a dome of 8-10 ft. inside diameter will have a minimum of 10 inches wall thickness, and a dome of 12 ft. diameter will have a minimum wall thickness of 12 inches.

Step 5) On the outer circle, fix a second post into the ground for the Height Compass. Attach the Height Compass chain and pull this chain tightly across the center of the dome to reach the inner circle. Attach a marker such as a steel ring to mark the length of the Height Compass. This length never changes for the whole dome.

Step 6): Pull the Center chain and the Height chain together to make sure that the two compass markers match. Now you have two chains marked: the Center Compass is the dome radius; the Height Compass is the dome diameter plus wall width.

Above: Step 4. A second circle is drawn using lime or chalk to represent the thickness of the dome wall.

Left: Step 5. The Height Compass post is fixed and the Height chain is pulled across the center of the dome to reach the inner circle.

Below left: Make sure that both compass markers (steel rings) match by putting your finger through both and pulling tight.

Below: A compass chain may also be fixed to a wooden post (a chair caster without the wheel provides a flexible connection).

Above: Students at Cal-Earth draw the dome section on the ground to decide on the springline (the dotted line at knee height).

Below: The Height Compass chain may be attached to the bag wall.

Deciding on the Springline With a Two Chain Compass

To set up the Height Compass, you will need to decide on the level of your springline. The "springline" is the level where the straight wall ends and the dome begins to curve. If your dome curves from the ground level, like an emergency shelter dome, then your springline is at the ground. But if you need more height inside your dome your springline will be higher up above grade, therefore you must build a straight stem wall before the dome starts to curve.

A good way to understand the size of your dome and where you would like the springline to be, is to draw it life-size on the ground and lie down inside.

When you have decided on the springline, the Height Compass is attached at that height. The most stable type of domes have their springline below grade level, because the earth becomes the buttress. If the springline is above grade additional buttressing and/or a tension ring may be needed.

The Height Compass can be attached to either a free standing post or directly to the dome entry wall in any way that allows it to turn freely. Examples range from a wooden post with a nail, a steel ring, even a chair caster, a pipe, a wire loop or stake between bags.

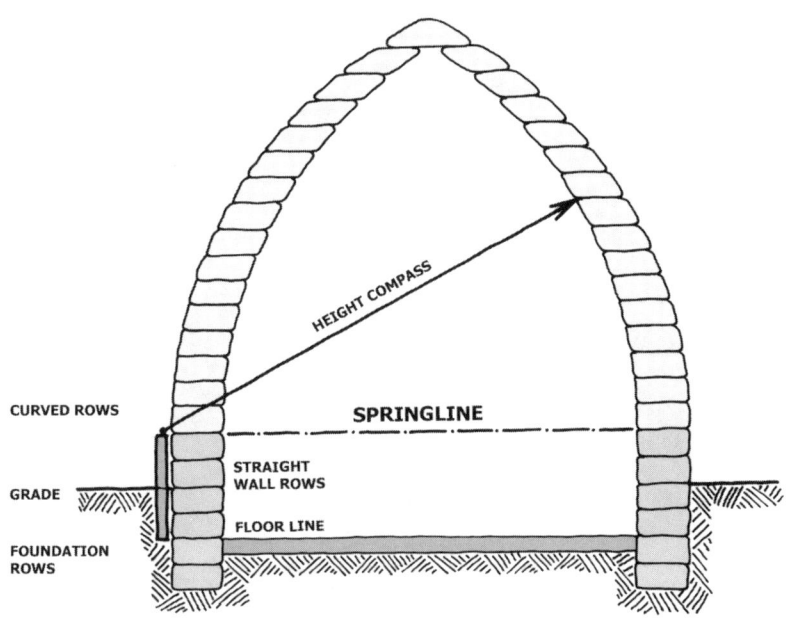

Above: Section of a dome. The curved dome rows are above the springline and the straight stem wall rows are below. If the Springline is above grade, the dome must be buttressed according to the engineering. (Small domes of up to 10 ft. diameter with a short stem wall may not need buttressing.)

Below: The Height Compass is located in a doorway. The chain must move freely up and down in the shape of the Lancet Arch. It's length measures from outside to inside the wall.

Above: A free standing Height Compass post is located just outside the doorway. An upside-down chair caster (without a wheel), is screwed onto a wooden post to attach the chain (or belt as shown here). This springline is two bags above grade, and the dome was buttressed by apses and seating areas.

Below: A metal Height Compass is located on the entry wall attached to the bags, with the springline just above grade.

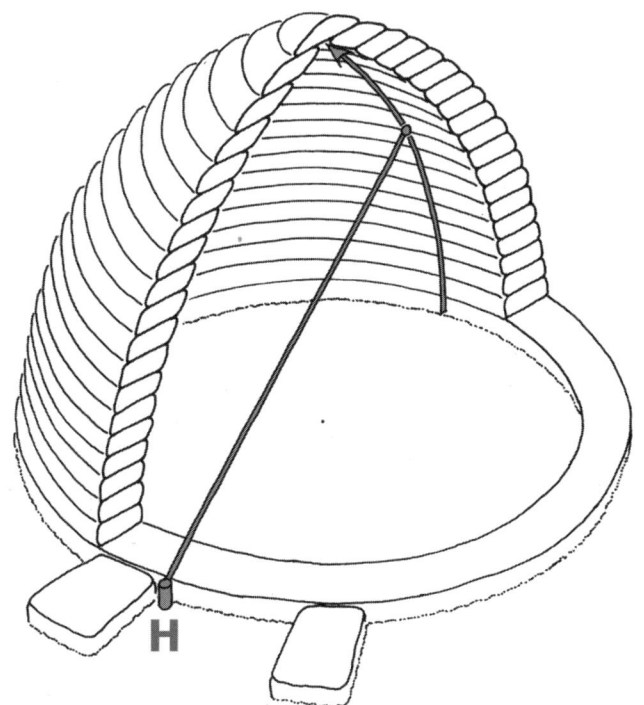

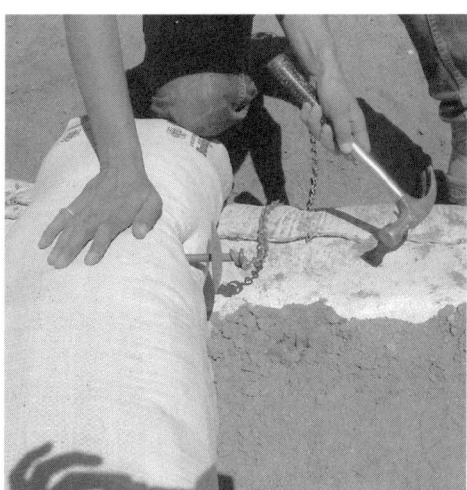

SUPERADOBE-SANDBAG SHELTER

Above and below: The two chains are pulled tight across the center of the dome. The springline for the Height Compass is at ground level.

Step 7) **Using the Two Chains.**

To build any Superadobe row, start building two feet of that row, on top of the last row, and parallel to it. Then holding both chains together, pull the Height Compass chain tightly over the dome center and hold it at the midline level of this next bag row. Next, pull the Center Compass up to match the Height Compass, and move the marker/steel ring along the Center chain to match the Height marker,

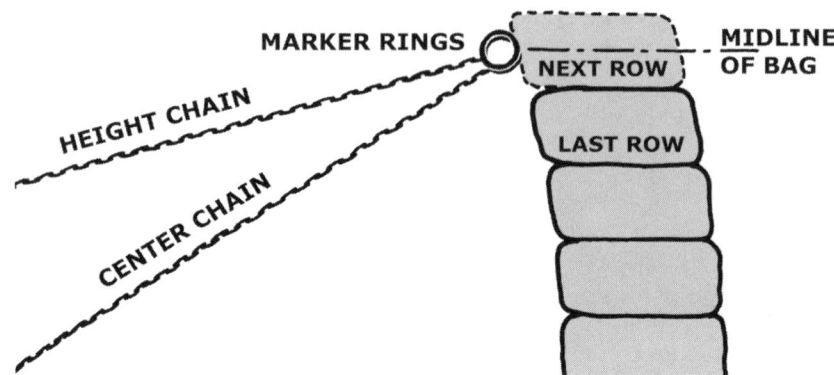

Right: Diagram showing the position of the two chains at Step 7.

Left and below: Both compass chains are pulled tightly together to determine where the next row of bags will be built. The Center Compass steel ring marker has been adjusted to match the Height Compass marker. It is being checked with a finger inside both rings.

so that you can put your finger equally inside both steel rings. Adjust this next bag row to meet the compass. Then drop the height chain, you will not need it until the next row, and continue working with the adjusted Center Compass.

So at the beginning of every row you will move the steel ring marker to make the Center Compass longer. It must get longer with every row you step up, otherwise your dome will become a hemisphere instead of a tall lancet/ogival dome.

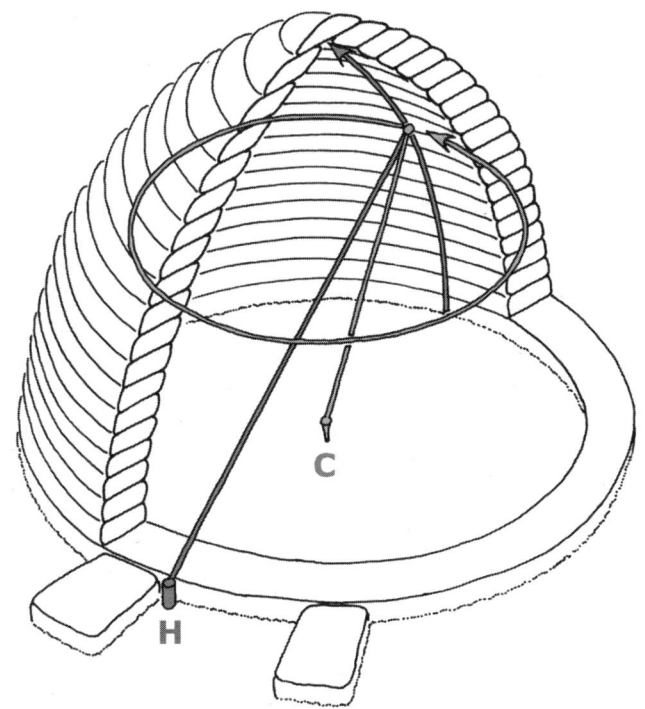

Left: Using both compass chains together to determine the length of the Center Compass at every row.

SUPERADOBE-SANDBAG SHELTER

Right: A finger or thumb is used inside the steel ring marker of the Center Compass and measure to the midline of the bag (not the top or bottom).
The bag position is measured both before and after tamping. Approximately one half inch is allowed for the bag to spread during tamping, depending on the earth mix and building technique.

Step 8) After you have set the length of the Center Compass, you will use it to build each row by measuring the circle of Superadobe bags. For a consistent measurement as you build, put your finger inside the steel ring marker, pulling the chain tight to touch the bag's midline (the fattest part of the bag).

When you have finished one ring of Superadobe using the Center Compass, and before starting the next ring you must make the compass longer again as shown in step 7.

The success of your dome depends on how closely you follow the compass. Because the bag spreads during compacting, you must allow for this when measuring. Whichever compass method is used, the resulting dome should be the lancet/ogival shape.

Earth-filled bags and a chain compass, by their nature, cannot create a perfectly accurate compass line. The "tolerance" is the distance allowed for the bag to be inaccurate. If the wall is wider, relative to the dome diameter more tolerance can be allowed, and if the wall is slim relative to the diameter, the tolerance is less. Therefore setting "half an inch tolerance" means we allow the bag to set within one half inch either side of the compass line.

Below: The movement of the Center Compass.

Below right: The ring marker measures to the midline of the bag.

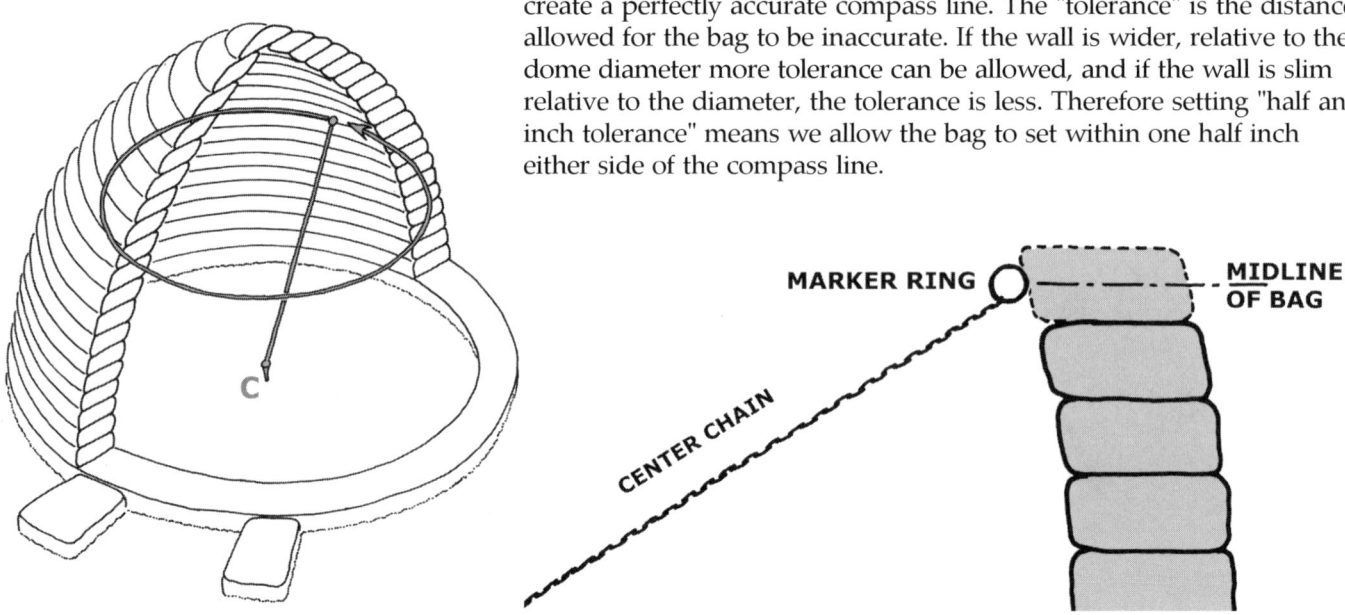

Left: The Center Compass can be set at any height, irrespective of the springline, since it gets its length from the Height Compass. Here it is shown at ground level. From the lowest row of the dome all the way to the upper most row, the compass is constantly used to measure the circle of bags for each row of the dome.

Below: For the user's convenience, both chains are made two feet longer than the dome diameter. The extra length of chain shown here can be tied to a counterweight and hung over the wall, so that it stays in place as the builder uses it.

Left: Measuring the upper rows of the dome is like all the other rows except that the compass chain must be pulled very tight to be accurate. Therefore a thin, lightweight chain should be used, which can be pulled taut.

Above: The marker which moves along the pipe with every row is a pipe clamp with an angle bracket. You can use a pre-marked pipe, or measure at every row.

Alternative Compass Designs

From a simple piece of chain to a sophisticated laser, a variety of compass designs can be the tool to build a dome. The Pipe Compass and the Sail Compass are two other compass methods we have used in our work.

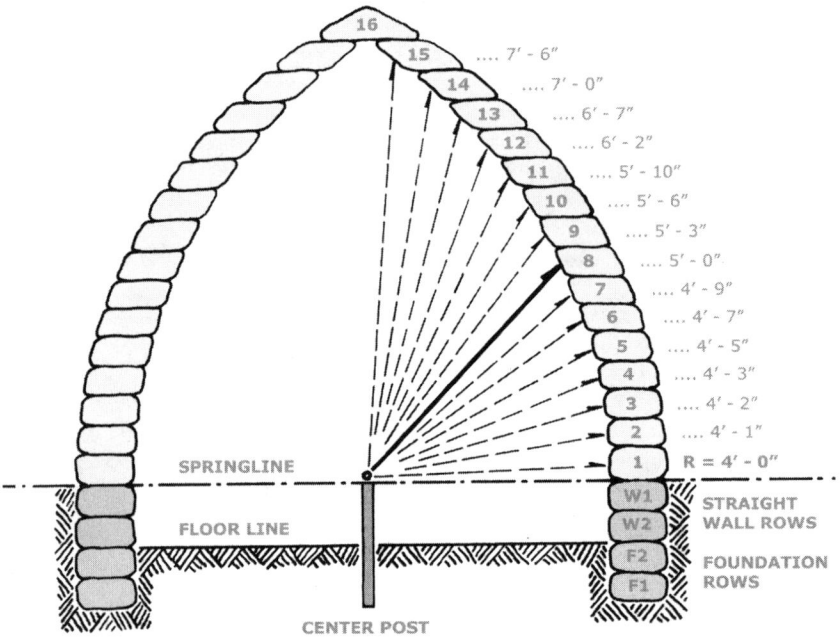

Above: A drawing of the measured pipe compass for an 8 ft. interior diameter dome, based on 6 inch high bag rows.

Above: At the center of the dome the pipe is attached to an upside down chair caster with the wheel removed so that it can turn freely. The chair caster is screwed to a wooden post which is set firmly into the ground and does not move until the dome is finished.

Right: A Pipe Compass does not need to be pulled tight, and can rest on the bag row as it is being built.

The Pipe Compass

For this method a length of fixed or telescopic pipe can be used instead of the two chains. Pre-mark the distances on the pipe which must be extended at each row of Superadobe. To determine how much the Pipe Compass needs to be extended at each row you need to draw a chart for the size of dome and height of bags you are building, unless you also use a Height chain.

For example, we could make this type of compass from a piece of pipe such as an extendible paint roller handle, a chair caster, a wooden post and screws, and a marker such as a pipe clamp with an angle bracket.

Each row is measured first on the drawing. However, this will only be accurate when the bag rows are a consistent predetermined thickness. For example, if we know that each row will come out 5 inches thick, we can pre-mark the compass length and row number, after which an untrained builder can use it more easily.

A variation on this theme is to install a vertical center pipe the same height as the top of the dome. A horizontal adjustable arm is attached to the vertical pipe and raised with every row. It is also shortened to describe the shorter radius of each succeeding row.

Above: A pipe compass being used.

Left and below: Exterior and interior of a completed Lancet dome showing a variation of the pipe compass which uses a vertical pipe in the center.

SUPERADOBE-SANDBAG SHELTER

The Sail Compass

For larger domes the entire lancet arch shape can be constructed with a compass using steel pipes or strong bamboo. This lancet arch "sail" compass rotates around the dome center, and has a wheel on the outer edge. At the dome center a short piece of steel pipe is cast into the floor slab. The vertical center pipe of the sail compass is slightly larger and slides over the pipe in the floor slab. This allows it to turn around 360 degrees.

To construct the compass shown here you will need 2 - inch diameter steel pipe, steel plate, a wheel and caster, and a welding facility, or appropriately sized bamboo. The curve of the compass can be drawn on the ground, and the compass is constructed flat on the ground before being raised into place.

Above, right, and below: The sail compass made for the Hesperia Museum and Nature Center. The pieces of pipe in the center of the horizontal bars allow it be dismantled.

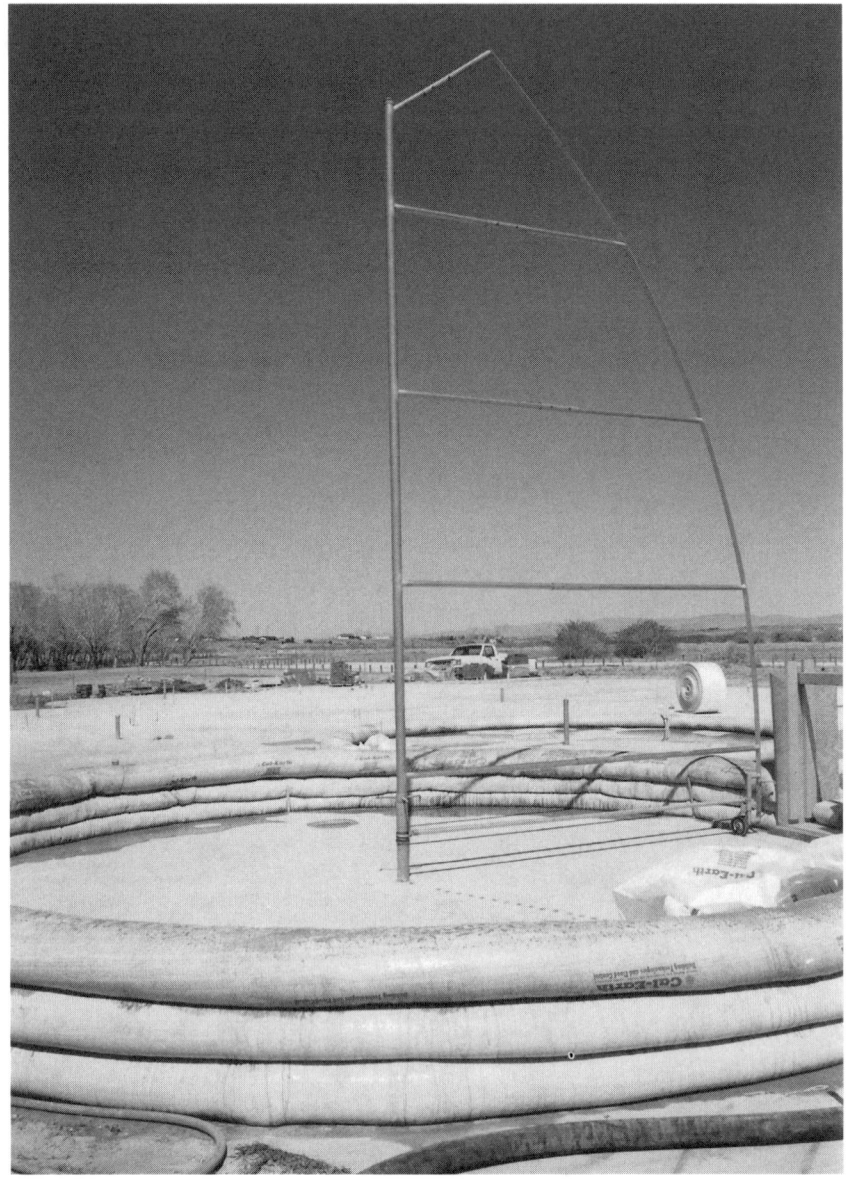

Left: The Sail Compass does not actually touch the bags, but a small extender arm is used, to avoid the chance of the bags interfering with the sail turning freely. The wheel in this case has been reinforced. We believe a better design would have placed the wheel further out. Window and door forms also need to be built so that they do not interfere with the free movement of the sail.

Theoretically, a Sail Compass could be built as both a scaffold and a compass by adding a second sail to make a wedge shaped unit. Planks could be laid across the wedge to make a scaffold with rails to conform to standard safety requirements for the builders.

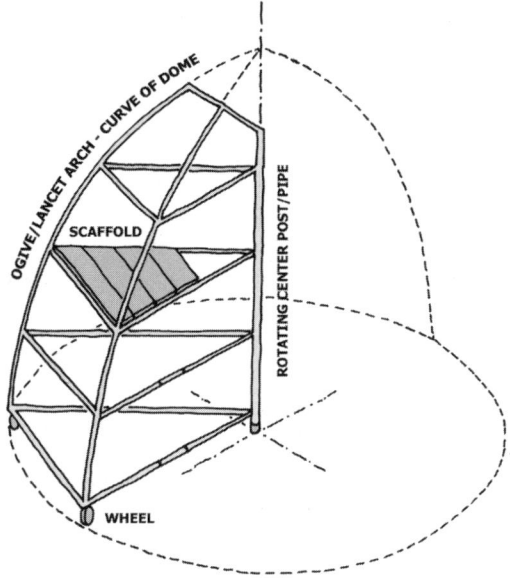

Left: Drawing of a Sail Compass concept for a combined compass and rotating scaffold.

Below: The Sail Compass and dome being constructed. When the dome is almost finished, the compass must be dismantled and removed (the circle is not part of the compass). It can be reused in another dome.

Conclusion

There are as many ways to form the Lancet arch as there is ingenuity in the human mind. The compass is the most important piece of your dome building equipment, from two chains and your skill, to a large Sail Compass. Without a compass we can lose our direction in building as in life. Many traditional Master Masons, whose long experience has attuned them to the forces of nature and gravity, have built beautiful, safe, domes and vaults without a compass. Their compass is already within them. For those of us who are now learning this art, the compass is our guide and friend.

"A Superadobe dome is like a child's toy made of concentric rings on top of eachother."

*all the precious words
you and I have exchanged
have found their way into the heart
of the universe*

-Rumi

GLOSSARY

adobe   A sun dried construction block made of earth-clay and sand; the essential building block of most traditional and contemporary earth structures.

apse   (see also **niche** or **pouch**) A half-dome. A curved, semicircular or polygonal recess into or projection from a building. In this architecure, the apse (architectural/structural term) is used to construct niches (spatial term) and pouches (Superadobe term).

arch   A curved structure, kept in balance by the pull of gravity, which supports its own weight. A geometrically formed symmetrical structure whose material spans over space by converting dead loads and live loads into compressive forces. Usually curved upwards toward the center.

arched entry   An arched door opening in an exterior or bearing wall. The arched entry in a double curvature compression shell, such as a dome, is projected forwards to form buttressed walls and vaulted roof.

asphalt   A dark-colored, solid bituminous substance composed mainly of hydrocarbon mixtures, naturally occuring in the earth of various regions. Also artificially formed as the byproduct of the petroleum industry. Liquefies when heated, for "hot-mop" application. Impervious to water.

asphalt emulsion   A fluid composite material consisting of asphalt emulsified with clay and water, for "cold application".

attachment   A material element sandwiched between layers of Superadobe bags onto which another structure or element can be attached.

bag material   Natural or man-made fiber woven into a tubular or individual flexible formwork into which the earthen filler material is poured. The bag material should withstand the pressure of filling and compacting, be a tight enough weave to effectively contain the material, be a loose enough weave to allow the material within it to cure and dry (see also **ultraviolet**).

barbed wire (4-point)   A double strand of zinc coated steel wire, having small pieces of sharply pointed wire (4 barbed points) twisted around them at intervals (of about 5 inches), used chiefly for fencing.

bed-womb   Inspired by the organic form of a womb. A small apse/niche built for storage, a bedroom, and more.

buttress   A wall built at an angle or thicker for shoring and bracing; a counteracting force that strengthens and supports, as in a row of arches. A projecting part of a wall built integrally therewith to provide lateral stability. Adding mass to the main structure to absorb outward thrust, which is developed in the main walls.

base Pertaining to Superadobe domes, that part of a domed structure which lies below the springline and above the foundation.

catenary A paraboloid curve; a catenary arch is in total compression, strengthened by gravity.

caterpillar A linear arrangement of shelters or pouches (see also **sinapsoapsis**).

cement, Portland cement A binder consisting of calcined mixtures of clay and limestone (burned to high temperatures in industrial kilns). Usually a grey or white colored powder.

coil A filled length of Superadobe tubing laid in a curve.

compass, dome compass A tool for describing and building circles and measuring distances. Used to create domes.

compressive strength The limit of resistance of a material to crushing under a weight measured in pounds per square inch.

corbel, corbelling To corbel, or corbelling, is a building technique in which a masonry, or other material unit, is made to step inward or outward from a vertical surface, gradually and incrementally; a corbelled dome is built of concentric rings of Superadobe coils, and is taller and more conical in shape than a hemispherical dome.

clustering A cellular design principle of interconnected domes, vaults, apses and other rooms, clustered together to make larger structures.

dome A curved roof similar in shape to an inverted bowl having a circular plan built over a circular or square room. A hemisphere or variation thereof, so constructed as to exert an equal thrust in all directions.

door vault A vault about the same width as the door opening which it buttresses.

eyebrow An architectural feature over the top of a door or window, similar in location and function to the human eyebrow over the eye; an additional mass or element on the exterior of an opening which sheds rain.

earth architecture Structures made of earth, mud, or clay and based on a philosophy of harmony with nature.

earth mix Earth used to fill the Superadobe bags. Earth composed of known constituents, such as a known proportion of water or clay. Used mostly for stabilized earth containing a known proportion of earth and a stabilizer.

entry vault A vault extending over the entry door, to buttress a dome and protect the entry from rain.

form, formwork (centering) A temporary structure of wood or metal that gives support to arches, domes, and vaults during construction.

foundation The basis of a building, the lowest division of a building wall usually partly or wholly below the surface of the ground.

key-stone, key-bag The wedge shaped piece at the summit of an arch, regarded as holding the other pieces in place.

lancet arch An arch having a pointed top.

layer A Superadobe layer is a single thickness of filled Super-adobe bag.

lime   A white or grayish-white, odorless, lumpy, very slightly water-soluble solid, CaO, that when combined with water forms calcium hydroxide (slaked lime), obtained from calcium carbonate, limestone, or oyster shells; used chiefly in mortars, plasters, cements and whitewash or paints.

mechanized Superadobe   A technique of pumping the earth mix into the long bags.

membrane, shell membrane   A curved surface that acts as a structural element, a single or double curvature compression shell, for example an egg shell.

mud-straw   A combination of mud and straw used to plaster adobe buildings, which improves stability, insulation, and waterproofing.

niche   (see also **apse** or **pouch**) An space recessed into a wall created by carving out part of a thick wall, or constructed by and apse or pouch. Small niches may be used as storage or display, larger ones can become entire rooms.

ogival arch   A pointed arch whose radius equals its span.

plaster   A spreadable, semi-solid, mixture of materials applied in a pasty form over walls and ceilings; a sacrificial layer of building material spread onto interior or exterior surfaces for protecting underlying structural materials. Types of plaster include: a) Stabilized earth plaster (a percentage of cement, or lime, or asphalt emulsion, is mixed with earth and water); b) Adobe plaster (clay-earth, sand and straw mixed with water); c) Gypsum plaster (sometimes mixed clay); d) Stucco (lime-sand-cement mixed with water); and other admixtures. Plasters may incorporate fibrous material such as straw, hair (horsehair) or coconut fiber for reinforcement.

pouch   (see also **apse** or **niche**) A bag-like cavity or sack. A dilation to form a receptacle, such as the saclike dilation of the cheeks of young gophers, or the receptacle for the young of marsupials. A Superadobe pouch is a protective shelter constructed as an apse or niche.

p.s.i.   "Pounds per square inch" - the unit used to measure compressive strength. Therefore 300 psi means that if a load of 300 lbs or less is placed on 1 sq. inch of stabilized earth it will not be crushed.

radius   The measure from center to circumference when describing the geometry of an arc, circle, arch. Used to define the distance of a dome or vault's wall/roof mass from its geometrical center.

rammed earth   A construction technique in which damp earth is rammed into a form to make a solid structure; it is especially good in damp climates.

reptiles   An innovative finish consisting of stabilized earth mud-balls, layered like tiles to mimic the patterns of fish scales and reptile skins. An exterior finish which solves the problem of cracks.

retaining wall   A wall for holding in place a mass of earth or the like, as at the edge of a terrace or excavation.

row   As in "Superadobe row". A horizontal layer of filled bag; used like the Masonry term "a row of bricks".

shell   A geometrically constructed figure, such as a dome or vault, with qualities of a continuous membrane under gravitational forces, due to its shape.

sinapsoapsis   A pattern by which a continuous coil of Superadobe can build many structures simultaneously, both by hand and by mechanized methods. Multiple apses which touch in a continuous, sinuous line built by the flexibility of the Superadobe coil.

skylight   A window in the roof of a building designed to bring natural light inside the structure.

soil bearing   The amount of pressure, usually in pounds per square foot (psf) or per square inch (psi), which the soil a can bear as weight from an object or building, before deforming or shifting,

spring line   The horizontal line or plane from which an arch, vault, or dome starts to curve.

stabilized earth   Earthen material for construction which contains binding additives to increase the compressive strength and/or moisture resistance. Earth bound together with any cementitious substance. Common stabilizers are cement, lime, ash, and asphalt (usually emulsified). Less common are cactus juices, enzymatic soil stabilizers, latex-based fluids, natural glutens and caseins, gypsum, and fire for producing ceramics.

Superadobe   Earth-filled sandbags of extremely long lengths to eliminate vertical joints. Sandbags, Earthbag, Superblock, Space block, are all the same.

tolerance   The permissible range of variation in the dimension of an object; an quantified, allowable inaccuracy.

UV ultraviolet   Ultraviolet light is beyond the violet in the spectrum and a part of sunlight. It has wavelengths shorter than 4,000 angstrom units. The UV rating factor for woven polypropylene bags indicates their resistance to ultraviolet light, in hours. For example 300 UV bags will begin to disintegrate after 300 hours of exposure to the ultraviolet light in sunlight.

vent   A smaller opening for air circulation.

vault   A single-curvature structure that creates a ceiling or roof in the form of a deep arch; built over long or rectangular rooms, halls, tunnels or other wholly or partially enclosed construction.

weep-hole   A hole for draining of accumulated moisture, as from condensation or seepage.

wind catcher   A structure that brings cool outside air into a building; the most effective wind catchers work in conjunction with basements, pools, and fountains to create evaporative cooling systems.

METRIC AND U.S. SYSTEM OF WEIGHTS AND MEASURES

1 millimeter = 0.0393 inch
1 centimeter = 0.393 inch
1 meter = 39.3 inches or 3.28 feet
1 kilometer = 0.621 mile = 3,280 feet
1 square centimeter = 0.154 square inch
1 square meter = 1,549 square inches
1 cubic centimeter = 0.061 cubic inch
1 cubic meter = 35.31 cubic feet
1 gram = 0.035 ounce
1 kilogram = 2.204 pounds
1 metric ton = 2,204 pounds
1 liter = 1.056 liquid quarts
Centigrade (Celsius) = °C   Fahrenheit = °F   °C = 5/9(°F-32)

CHRONOLOGY
SUPERADOBE AND CERAMIC HOUSES
R & D, TESTS, PERMITS

Earth and Ceramic Architecture and their construction technologies, have been tested at Cal-Earth Institute since 1991, for the California Building Codes. The building systems here are developed as the spin off from Khalili's presentations at NASA'a first symposium on "Lunar Bases and Space Activities of the 21st Century" in 1984 and subsequent conferences. The structural design and test procedures of the model houses and Hesperia Museum are based on the engineering design and supervision of P. J. Vittore Ltd. Engineering, Arlington Heights, IL. Permits and documentation concerning the static and dynamic load tests on the Superadobe and masonry prototype structures are on record in the City of Hesperia, California.

1976-1979	Research and development of the Ceramic Houses technology using Geltaftan (earth and fire).
1979-1984	Application of the Ceramic Houses technology for village housing and a school in Iran, described in the books "Ceramic Houses and Earth Architecture" and "Racing Alone". Exhibitions in the Centre Georges Pompidou, Paris and other institutions, as well as publication in many journals and magazines. The CCAIA (California Council of the American Institute of Architects) gives the award for "Excellence in Technology".
1983-1987	Khalili begins teaching at SCI-Arc (Southern California Institute of Architecture). R & D in Ceramic Houses continues in the United States with grants from the National Endowment for the Arts and Kit Tremaine. Projects such as "Future City-Villages" develop. Awards include a "Special Recognition" from the U.N. International Year of Shelter for the Homeless, and "Housing for the Homeless" from HUD (U.S. Dept. of Housing and Urban Development).
1984-1990	"Magma, Ceramic, and Fused Adobe Structures Generated In Situ " is presented at NASA's first symposium on Lunar Bases and Space Activities of the 21st Century, held at the National Academy of Sciences in Washington, D.C, and is later published by the Planetary Institute, Houston. Khalili gives a colloquium at Los Alamos National Laboratory in New Mexico, on their invitation, and lectures at the Physics Auditorium to a mixed public and inhouse audience. He also lectures at MIT and Princeton University amongst others, and conducts research with the Space Studies Institute of Princeton, New Jersey, and McDonnell Douglas Space Systems, in California. The paper "Lunar Structures Generated and Shielded with On-site Materials" is published in the Aerospace Engineering Journal of ASCE (American Society of Civil Engineers), and is awarded best paper for that year. See "Ceramic Houses and Earth Architecture" for a more extensive chronology until 1990.
1991-1993	Cal-Earth Institute is established in Hesperia, California. Construction of full scale prototypes begins. R & D into the Superadobe (sandbag and barbed wire) technology is based on the lunar proposal "Velcro-adobe" and "flexible containers".

1992	Plans and engineering for a 20 ft. diameter masonry dome (Rumi dome), and a 15-18 ft. diameter Superadobe dome are reviewed by Hesperia Building and Safety Department in consultation with I.C.B.O. (International Conference of Building Officials).
1993	Engineering review demands liveload tests to simulate seismic, snow, and wind loads. Static eccentric loading of full scale domed structures to 200% of code requirements was performed and monitored by independent engineers Inland Engineering Corporation -no deflections observed.
1994	U.B.C. categories M-1 and B-2 structures are allowed in Hesperia as part of a prototype program. UNDP/UNHCR commissions fifteen Superadobe domes which are built by the refugees themsleves in Khuzestan province, Iran after "Desert Storm" , the Kuwait-Iraq war. Plans and engineering for the Hesperia Museum and Nature Center, consisting of 14 Superadobe domes of 20-50 ft. diameter, 2 vaults, and a ceramic tower are reviewed by Hesperia Building and Safety and the ICBO.
1995	Simulated dynamic and static load tests are performed to establish the safety for all UBC occupancy categories on full scale prototypes of both dome and vault typologies: a) Superadobe dome (double curvature compression shell), b) vaults (single curvature compression shell), and c) masonry dome. Testing is monitored by the ICBO approved laboratory, Southwest Inspection and Testing who report that, "All tests have exceeded the ICBO and City of Hesperia requirements".
1996	Construction permits are issued by Hesperia Building Dept. for the Hesperia Museum and Nature Center, as reviewed by the ICBO. Construction begins.
1998	The Superadobe 3- and 4- bedroom model house plans "Earth One" A and "Earth One" D are reviewed and permitted by Hesperia Building Dept. as stock plans. Encompass Learning Center is designed and residential domes are permitted by Nevada City Building Dept., California, in consultation with ICBO. Twelve double dome units are constructed.
2000	Plans and engineering for the "Earth One" stock plans are updated for the 1997 UBC and approved by Hesperia Building Dept. Construction begins on the "Earth One" model house at Cal-Earth Institute, Hesperia, California.
2002	Holy Resurrection Monastery's twelve Eco-Domes are approved as monastic residences by San Bernardino County, California. Construction begins.
2004-2005	Plan review for the Eco-Dome stock plan (guest house/efficiency unit) results in permitting by Hesperia City Building Dept., and the double Eco-Dome (800 sq. ft. residence) stock plan is permitted by San Bernardino County Building and Safety Dept.
2005	The Aga Khan award for architecture goes to Nader Khalili, for "Sandbag Shelter Prototypes". Cal-Earth Pakistan is established, building and training the victims of the Central Asian earthquake on a large scale.
2005-2007	Eco-Domes are permitted and built in several California jurisdictions. The Superadobe domed and vaulted plans used globally to construct projects. The Hesperia City and San Bernardino County stock plans are kept updated for the UBC and California amendments to 2007, the time of writing.
1994-2007	Many hundreds of large and small scale houses, schools, orphanages, and other building types, have been constructed globally.

INDEX

A

adobe 14
   Velcro-adobe 15
apprenticeship 17
apse 90, 150, 180
arch 60
   dry-pack 260
   brick 264
   geometry 258
   leaning 80, 186. *See also* vault
   long bag 117
   principles 257
   small bags 260
   Superadobe 259
   varieties 109, 164
asphalt 135
attachments 85

B

bag. *See* Superadobe: bags, tubing
   cutting 37
   ending 152. *See also* corner
   filling 37, 47, 73, 119. *See also* tube
   level 46, 58
   removing 189
   small 270
   tamping 74
   types 26
   UV rating 26
   width 150
barbed
   wire 15, 17, 27, 42, 67, 76, 151, 270, 274
   extensions 153
   handling 155
   placing 43
building
   department 272
   permits 272
buttress 50, 83, 183, 243, 275

C

coils 14
compacting. *See also* tamping
compass 27, 73, 108, 113, 118, 152, 279
   center 39, 281, 288
   height 281, 286
   pipe 291
   sail 292
   two chain 282
compression 14
corbelling 16, 72, 186
   free-style 166
corner 116. *See also* window
courtyard 124. *See also* design: concepts: courtyard

D

design 88
   concepts
      clustering 89
      wall 90
   variations 88
disaster 9, 13, 243
dome
   geometry 33, 52. *See also* compass
   top 133
   upper rows 72
door 50, 51, 68
   arch 60, 110. *See also* window
   carpet, fabric 147
   vault 80. *See also* entry

E

earth
   stabilized, unstabilized 244, 270, 275
   tests 28
engineering 241, 272
   compression and tension 241
   domes and vaults 241
   stresses 244

entry   152
  apse   186
  flood control   121, 147
  "mineshaft"   134
  shell   186
  straight wall   147
environment   13
eyebrow   62, 111, 156
  seashell   158
  window. *See also* window

F

finish. *See also* plaster
finishes
  sponge   191
fireplace   151
flexibility
  design   9
  Superadobe   47, 112
forms   110
  arched   97
foundation   35, 97, 150
  compacting   36
  level   35, 36
  rows   37
  subsoil   32
  trench   35, 36
fountain   208
friction   15

G

geometry   16
global warming   13
gravity   13
gutter   191

I

infrastructure   14

L

lintel   134
loading   246. *See also* engineering
  seismic   244

M

mechanized construction   144
"Mineshaft" entry   134
moon   15

N

niche   185

O

openings   98, 138. *See also* window, door
  barbed wire   67
  size   67

P

pipe
  window   56
plan
  orientation   32
plaster   114
  finish   104, 123
  interior   168
  minimum   190
  mud-straw   169
  rough   100
  smooth   121
pond   208
pouches   90, 96. *See also* apse
practicing   17
principles   16

R

reinforcement   15
Reptiles   135. *See also* plaster
  geometry   140

S

sandbag   276
  standard   268
sandbags   14
scaffold   113
seashell   13, 180
seating   102
shelter
  temporary   28
Sinapsoapsis   90
skylight   133
springline   284
squinch   146
structural. *See* engineering
  shell   241
stucco   124
summary   19

Superadobe   14, 16
  bags, tubing   26, 27, 47, 276
  patent   10, 15
symmetry   274

T

tamping   40. *See also* compacting
tension   14
tests   246
  arches   263
  perfomance   274
  seismic   246, 272
torch   189
trowel   174
tube
  lightweight   119

V

vault   80, 163
  entry, wide   84
ventilation   58, 111, 129
vents   130

W

wall   90
  connections   154
  stem   120. *See also* base
waterproofing   85, 172, 190. *See also* asphalt
  asphalt   135
  quick   85
  temporary   85
window
  arch   60, 110
  circular   182
  eyebrow   62, 156. *See also* eyebrow
  form   60, 64. *See also* forms
  pipe   56, 97, 128, 130, 138
windscoop, wind-catcher   133